W9-ARL-719

# MILLVILLE GLASS:

*The Early Days*

# MILLVILLE GLASS:

## *The Early Days*

*by*

VIRGIL S. JOHNSON

DELAWARE BAY TRADING CO., INC.
*Millville, New Jersey*

*Library of Congress Catalog Card Number: 79–155302*

ISBN: 0–9600330–0–9

Manufactured in the United States of America by
The Haddon Craftsmen, Scranton, Pa.

Dedicated to the Glassworkers of Millville—men who have helped to build a community which has progressed and prospered in the 170 years of its existence.

# Acknowledgements

The study of the glass industry in its progress and development in South Jersey is one of deep and absorbing interest. The general public has had no clear conception of the practices in vogue in the early history of the glass trade in this section of the state. It is the design and purpose of this study to examine the conditions of this earlier period and contrast them with the conditions as they exist at the present time.

The staple commodities existing in South Jersey, viz., the ample supply and superior quality of the glass sand found in immense beds in nearly all the counties, and the apparently inexhaustible supply of wood for consumption in the furnaces, led to the selection of this section of the state for the establishment of glass factories—factories which for many years multiplied with extraordinary rapidity. In several instances they were built absolutely in the woods, often ten to fifteen miles from any town or village.

Rude houses were erected to shelter the workmen and their families; these were scantily furnished and unattractive in their surroundings. In these squalid quarters, remote from schools and churches, with the boys at too early an age put to work in the factories, girls and boys alike growing up in ignorance and without the softening influence of rational amusements and the restraints of Christian guidance, the existence of a glassworker was one of frustration. However, men held firm and marched toward a better day.

In making a study of the history of the glass industry in Millville every possible precaution has been taken to make the information conveyed both authentic and accurate. The information has been assembled by me over the 56 years I have been a newspaper reporter and those to whom I would like to give credit for helping me are:

*William McCarthy* who had long experience both as a skilled glassworker and a factory manager—a man who worked with *Ralph Barber* who made the famous *Yellow Rose* paperweight.

*Foster V. Mitchell* who has been helpful in checking my copy to eliminate errors.

*John Mulford*, a South Millville resident, who helped make the world's largest bottle—capacity 108 gallons—at the South Millville Plant of the Whitall-Tatum Co. for exhibit at the St. Louis World's Fair.

Former Governor *Edward Casper Stokes* and the late Millville Mayor *George W. Payne* who were principally responsible for the *Cash Bill* in the State Legislature.

*Thomas Eames* former Secretary of the GBBA International Office, resident of this city.

*Joseph Smith* a skilled machinist and a past president of the Central Labor Union of Cumberland County.

*Charles Pepper*, an expert glassblower.

*Herbert Vanaman*, of Port Elizabeth, N.J., who has assembled a considerable amount of information about glass factories in Cumberland County.

*Charles Pennington*, who gave me photographs of glassworkers.

*George S. Bacon*, Manager of the *Whitall-Tatum* factories for many years.

*Frank H. Wheaton, Frank H. Wheaton, Jr.* and *Martin G. Weber* who have kindly provided information about *Wheaton Industries*.

*Roger F. Scott*, Manager of the Millville Plant of the *Kerr Glass Manufacturing Co.*

With their assistance I have assembled this history which may be of some value to those who are interested in the glass container business and who would like to know how the industry progressed through the years from 1806 to the present day.

Millville has always been a glass town. Men have earned their livings making glass: window glass and glass containers; fancy ware and laboratory products. For over 164 years glass has been in Millville. And today the industry is expanding more rapidly than ever with more than 5000 people employed in Millville glass plants.

VIRGIL S. JOHNSON
*Millville, N.J.*

# Contents

# MILLVILLE GLASS:

## *The Early Days*

## CHAPTER 1

# South Jersey's First Glass Furnaces

THE Township of Millville was set off from Fairfield and Maurice River by an Act of the State Legislature passed February 14, 1801, to become effective in March, 1802.

At that time the township included what was later known as Landis Township, which was incorporated with Vineland by a vote of the people in 1952. Landis Township was set off from Millville Township in 1864, when Charles K. Landis purchased land from Richard D. Wood of Millville and Philadelphia to establish Vineland and the Township named after himself.

When Millville was organized as a city through an Act of Legislature approved February 26, to take effect March 1, 1866, it became the largest city in area in New Jersey and remained in that position until Estell Manor was incorporated as a city. Vineland Borough, consoli-

An early veiw of Millville from the West Bank of the Maurice River. Looking toward the Maurice River Bridge and the early "center city" section.

dated with Landis Township in 1952 is now the largest city in area.

Prior to 1800, the tiny settlement of homes located near the Maurice River had been called by various names—Shingle Landing, The Bridge, and Maurice River Bridge.

History relates that one Lucas Peterson built a house on the west shore of the river around 1755. It later became a tavern and was finally destroyed by fire. This place is believed to have been owned, at one time, by

Alexander T. Moore, formerly of Bridgeton. The tavern was kept by Philip Souder, Jr., in 1793, and by Benoni Dare in 1796. Moore built an addition to the house in 1796 for a kitchen for the use of his tenant.

There was said to have been a saw mill built within the bounds of what is now Millville, called Leamings Mill.

Millville is believed to have derived its name from the mills that were located in the area.

Col. Joseph Buck, one-time sheriff of Cumberland County, a resident of Bridgeton, with friends, had enough faith in the area east of the Maurice River to buy a considerable tract of land, about 1795, and divide it up into building lots which he offered for sale. In spite of the sandy type of the soil, he was successful in selling home sites in the section just east of the river. He built a home for himself on the northeast corner of High and Main Streets. It was the most pretentious dwelling in Millville and was erected in 1796.

The lot on the southeast corner of High and Main Streets was the first one sold. The 60-rod plot was bought by James Sweatman on May 10, 1797, but was soon sold by him at a nice profit. Nathan Leake, a native of Deerfield, bought an acre plot on the northwest corner on which the Millville Office of the South Jersey National Bank now stands. The development of Millville continued, but it was slow.

Glass may have been made as far back as 2500 B. C. Some researchers say glass beads were made in Mesopotamia about that time. Glass walls in Roman baths have been mentioned, and some historians say glass was being made in Venice in 1268 B.C. But in America, the earliest mention of glass making referred to "a glass house being erected half a mile from Jamestown, Virginia Colony" that was operated with "alacrity and success" early in 1609. But it didn't last.

Our interest, however, is in South Jersey, and, in particular, Millville.

Stories have been told of an early glass factory along the Mullica River, and another in the Pines of New Jersey, prior to 1776, with the report that glass containers were made there. The information about them, though, is vague.

There is proof in the New Jersey Historical Society's records that Casper Wistar, born in Germany, came to America in September, 1717, and that on January 7, 1738, he began buying large tracts of timberland in West Jersey. He and his son Richard owned 2000 acres, some two miles from Allowaystown in Salem County. Some of the acreage was cleared. Rough sheds, called factories, a general store, cabins for workers, and a mansion house for Wistar were erected. It is difficult to find any indication of South Jersey's first glass works today.

Wistar knew where to obtain skilled glass workers. He sent to Belgium for four expert glassworkers. Casper Halter, Martin Halter, John Wentzell, and Simon Greismeyer came over with the agreement that all expenses would be paid by Wistar. He agreed to provide homes, food, fuel, and helpers. They were also to receive one-third of the profits of the production; for this they were to teach Wistar and son Richard the secret art of glassmaking.

Wistar was smart and he was ambitious. In the 22 years he had been in America he had become rich. Nine years after he arrived he married Miss Catharine Johnson of Germantown, a lady of the Quaker faith. He joined the Society of Friends in 1725 and soon became prominent in Philadelphia. He gained fame and prosperity from the manufacture of Philadelphia brass buttons, which he guaranteed would last seven years.

Having come from Germany where there were expert glass makers may have been a reason why he decided to build a glass factory. There weren't any here—hadn't been for more than half a century. Evert Duycking had tried to make glass in New York, and Jacob Melyer, who succeeded him, operated the factory for a time, but a search of the records indicates that reliable glassware, containers and window glass, were imported in 1739 when Wistar decided there was profit to be made in the manufacture of window glass and bottles.

He chose a Salem County site for the factory. Why, not anybody can say. He bought 39 acres along a branch of Alloway Creek, a mile above what was known as Thompson's Bridge, on January 7, 1728, and by the summer of 1739 he had acquired 2,000 acres of land through which the turnpike from Salem to Pilesgrove ran. The factory was built by the end of 1739. Glass making started late that year. It is believed that Wistar chose the site because of its nearness to Philadelphia, the sand, the wood, and the transportation facilities.

When Casper Wistar built that first glass-making fac-

T. C. Wheaton Glass Plant ca. 1900. *Granville Thomas photo.*

Whitall-Tatum Co. glass works at Shetterville (South Millville) ca. 1900. *Chas. B. Pennington photo.*

tory he also created a village that included ten homes for the workers, a large mansion house containing six rooms on a floor with a bake house and wash house, and a convenient storehouse where a well assorted retail shop had been kept for thirty years and was said to be as good a stand for the sale of goods as any in the county.

Most of the early factories for making glass were built in the woods because of the nearness to large tracts of woodland because wood was used in the furnaces. Therefore, it appeared necessary to have homes for the workers and a store where the glassworkers spent most, and in many instances, all of their earnings. It was not unusual for many of the employees to be indebted to the store when the factories closed for the summer stop.

In later years, when glass plants were established in cities where there were stores, the companies set up stores of their own and insisted that their employees deal at the company store. In the old days there were many employees of factories who were always indebted to the company.

The store provided everything—home, coal and wood, food, clothing, shoes, the doctor, and the worker could obtain cash by calling at the office.

James Lee, a promoter, who came out of the Kensington section of Philadelphia, was anxious to build glass factories in South Jersey. However, he was a promoter and he was willing to build anything. His name first appears in a history of Port Elizabeth where James Lee was said to have built what was known as the Eagle Glass Works prior to May 23, 1799, on leased ground. On January 29, 1805, James Lee bought 127 acres of land in Port Elizabeth from Nathan Hand. He later disposed of his interests in Port Elizabeth and established the first glass making factory near what is now Buck and Mulberry Streets, Millville in 1806. His Millville venture, however, remained in operation, and while there were lean years, the project was kept alive and expanded. James Lee did not remain in Millville very long. In fact, the records indicate he did not remain anywhere very long.

The factory he established was for the manufacture of window glass. It passed into control of Gideon Scull, and then Nathaniel Solomon, manager for a company of blowers who failed to make a success of the venture. The business was then acquired by Burgin, Wood and Pearsoll, who in turn sold to Scattergood, Haverstack & Co., and they sold to Whitall and Brother in 1834. In 1849 the firm name was changed to Whitall Brother and Company, and in 1857 to Whitall-Tatum & Company, which remained the title of the company until 1938 when the Armstrong Cork Company of Lancaster, Pennsylvania, purchased the upper and lower plants of the Whitall-Tatum Company. They sold to Kerr Glass Manufacturing Co. of California in 1969.

It isn't clear when window glass was replaced by the manufacture of bottles.

The Stanger Brothers were conspicuous in the development of the glass making industry in the Glassboro area. One of them had been with Richard Wistar at his Alloway factory. He was then known as Jacob Stenger. He was 18, and he, with John Kendiel, 17, evidently ran away from the Wistar settlement. As related elsewhere, there was an advertisement in the *Pennsylvania Chronical and Universal Advertiser* on April 18, 1770, offering a $20 reward for "two German servant lads, Jacob Stenger and John Kendiel," and giving a description of both.

Jacob Stenger, or Stanger, as the family was later called, was one of the Stanger brothers who started the second glass works in New Jersey in the Glassboro area about 1780–81. In 1783, the interests of three of the Stanger brothers had been acquired by Thomas Carpenter and Samuel Tonkin, and, later, Col. Thomas Heston purchased one-quarter interest of Solomon Stanger. In 1786, the firm became Heston and Carpenter. Heston died in 1802. His widow and Carpenter continued to operate the factory until 1808, when Edward Carpenter acquired his father's interest, and Peter Wickoff that of Mrs. Heston. The firm name changed to Edward Carpenter and Company. Carpenter, with different partners, continued to operate the plant until he died in 1813 and the factory became idle. It was later revived under new ownership.

In the meantime, a number of workmen under leadership of some of the Stangers built a new factory some 400 yards south of their first plant and operated it under the firm name of Rink, Stanger and Company. This was the Olive Glass Works, and in 1824, Jeremiah Foster bought it and merged the business with the Harmony Works. Eventually, the plant became known as the Whitney Glass Company. In 1918 it was purchased by the Owens Bottle Company and it continued to operate for a number of years.

CHAPTER 2

# Glass-Making in its Infancy

THE location of glass container factories in South Jersey began at a very early date, more than 160 years ago. The first furnace was established at Allowaystown, in Salem County, and the second at Clementon, Salem County. Then followed Atco, Winslow, Waterford, Williamstown, Port Elizabeth, Millville, Bridgeton, Salem, Quinton, Clayton, Malaga, Glassboro, Woodbury, Fairton, Vineland, and Minotola. The most flourishing of these factories were those located where the advantages of water transportation were available, thus greatly cheapening the outgoing and incoming freightage. Millville, Bridgeton, and Salem especially profited by this means.

The glass companies built vessels to carry materials to the factories, and convey the products of the factories to market. Millville had several vessels plying between that port and Philadelphia, and finally had built two large steamers, one of which carried freight for the company to and from Philadelphia, while the other made weekly trips to and from New York; these steamers were also licensed to carry passengers, and many persons enjoyed trips on them.

For many years all these factories enjoyed a high degree of prosperity and proved to be extremely lucrative to their owners, it being no uncommon occurrence for a member of a firm to retire with an ample fortune; but with the growing scarcity of wood and the increased cost of freighting materials and manufactured goods, many of these factories, remote from railroads, ceased to be profitable and were abandoned.

Gradually, conveying of freight by water was abandoned, for the railroads ran spurs into the factory yards and repeatedly made large reductions in freight charges. Today, much of the shipping is done by motor transport.

The long strikes at Clayton and Bridgeton may be said to have most clearly demonstrated the fact that the old methods were doomed, and that the final adjustment of their difficulties was to mark the dawn of a new era in the glass industry. The result worked out was the unionizing of nearly all the factories in South Jersey; one after another, Glassboro, Clayton, Bridgeton, Elmer, and Fairton united in agreement to union rules; the power of the Glass Bottle Blowers' Association was firmly established, and, better for all, an agreement was effected between the association and the manufacturers. This was followed by an advance in the prosperity of the glass trade far beyond anything it had enjoyed during former years.

The Upper Works, or Glasstown plant of the Whitall-Tatum Co., looking north toward Union Lake. The R. D. Wood Company's mill is marked "X." Exact date unknown. Probably prior to 1900. *Virgil Johnson photo.*

View of the Upper Works of the Whitall-Tatum Co. ca. 1910. *Virgil Johnson photo.*

This happy result can only be attributed to the wisdom and conservatism of the officers of the association, who held the men from violent outbreaks and convinced them that no favorable outcome could be looked for if they resorted to rioting and the destruction of property.

In the earlier history of the glass industry the glassblowers were entirely without organization. The men met the manufacturers in their individual capacity; each man took to his employer his special complaint and stated his personal grievance. There was no fixed list to which all the blowers adhered; there were great differences in the various factories in the list prices, and even in the same factory, men working on identical ware were being paid different prices per gross. There was no cohesion, no thorough understanding, and, consequently, the blowers suffered. Workmen could not deal with employers in this manner; the interest of one as to the time employed, the character of the work to be done, the price to be paid for labor, was the interest of all. The glassblowers were to learn this fact from experience, and so it came about that the various groups of blowers at the factories selected committees of their most judicious and conservative men and appointed them to stand between the men and the employers and represent the interest of their fellows. The wisdom of this course was at once apparent: it conserved the rights of the men; there was a more clear and comprehensive statement of differences; difficulties were settled without clamor or wrangling, and it was more satisfactory to the employers to deal with their men collectively through the medium of a chosen few, rather than with each one separately.

Thus was instituted a better order of things. The existence of this incipient organization gave the men character and standing in the eyes of the employers, and greatly increased their own self respect. It gave the blowers a grip on the restless and disorderly of their own number, and led them, for the first time, to recognize the fact that the betterment of their condition was largely in their own control.

Out of these committees, naturally, grew the Glassblowers' League, where matters of interest to the craft could be discussed, movements inaugurated toward the establishment of a uniform price list, and the committees instructed for demands upon or concessions to the manufacturers.

The next step was the creation of the National Association, with provision for an annual convention composed of delegates from all existent union glassblowers' leagues. The business of this convention was to discuss the various items of the list and adjust prices, to hear and determine complaints and appeals, to elect officers to supervise and execute the affairs of the association *ad interim*.

This association has grown to be a very powerful organization, probably as thorough and complete in its management as any labor organization in the country. The officers are carefully selected and are cautious and conservative men who recognize the folly of violent outbreaks and give their attention to winning the desired results by diplomatic methods. The President, Vice President, and Executive Board are constantly on the move from point to point to settle differences that arise, give counsel in case of difficulties arising between men and employers, and advancing in every possible way the best interests of the craft.

Regularly the manufacturers have their meeting, and the Glassblowers' Association has its convention. If any differences have developed, at the conference these respective organizations discuss the same and instruct their representatives as to their views in the matter in dispute. Negotiations for a new contract covering wages and working conditions are held as each existing contract nears its expiration. Recently these contracts have been for three year periods.

These conferences are conducted in a businesslike manner. Each side has respect for the economic strength and ability of the other; such are the improved conditions under the organization as established and perfected for these later days. It is an ample illustration of the old axiom "In Union there is Strength."

This illustration is of the methods of the Glass Bottle Blowers' Association of the United States, and will apply, also, to the American Flint Glass Workers' Union.

## CHAPTER 3

# Company Houses

THE houses furnished by the company, which the men occupied, were miserable affairs without convenience or comfort. They were mostly two rooms with a shed kitchen on the first floor, two sleeping rooms on the second floor, and a large attic, generally utilized as a sleeping room for the children. There were no adornments about the dwellings, and the yards were in rough and unclean condition. Many of these houses remained for a number of years in Millville, Bridgeton, Williamstown, Winslow, and South Glassboro. Millville had its "Grumble Alley". The rents demanded were about on a par with those of six or seven room houses in the immediate neighborhood which were kept in good repair and the yards of which were in prime condition. The effect of these houses and their surroundings upon the social and domestic life of the workmen and their families was very bad. The influence of this kind of life was demoralizing; it tended to shiftlessness and improvidence; caused the families to be extravagant and careless of appearances; it demoralized the domestic life, and led the men, particularly, to regard the home as a mere accommodation for eating and sleeping, and the hotel and saloon as the place to spend all spare time; in consequence, the wages of the men were largely spent at these places, and but little was saved.

These conditions were radically changed during the next two decades; the company houses were greatly improved, and in most cases rents were reduced to correspond with rents as charged by private parties owning houses in the same neighborhood. But the marked improvement was in the fact that so many of the glassblowers, as a result of their own industry and economy, came to own their own homes, and some of them were fine dwellings.

In 1860, in the City of Millville, only eight glassblowers owned the property they occupied; in 1900, 43 percent of these men were property owners. Third Street, almost its entire length on both sides, was studded with homes of glassblowers; these houses were of fine appearance; had seven or eight rooms, and varied in cost from $1,500 to $2,500. The lots were large and were adorned with plants and shrubbery. The dwellings were handsomely finished and presented fine pictures of comfortable home life. One of the most attractive houses in the city, very elegantly furnished and located on Broad Street, was owned and occupied by a glassblower, and was the result of his own, unaided labor and strict economy. While this was true of the City of Millville, which was described at this time as the model glass manufacturing city of the county, it was not so marked in the towns of Bridgeton, Salem, Clayton, or Glassboro; the percentage of property owned was much smaller.

In the earlier days of the glass industry, one reason for the fact that but few glassblowers acquired property was in the uncertainty of tenure. Many of them, after working out a fire in one locality, preferred to seek another place for the ensuing season; one glassblower in Millville, after working in nineteen different factories, went back to work in his old place. The manufacturers, too, were often arbitrary, and in some cases would make a total clearance of shops, turning out the old hands and putting on entirely new men. Under these conditions the glassblowers were rolling stones and acquired no fixed habitation to dignify by the name of home. This state of affairs changed later, and a better and kindlier feeling between employer and employee was established promising much good for the future.

## CHAPTER 4

# The Company Store

THE company store, long since out of business in South Jersey, came into existence with the establishment of the first glass-making factory in the United States. Casper Wistar opened the first company store.

As we will later see, the company stores were always a source of serious difficulties, heated discussions and a vast amount of legislation.

The stores were coincident with the establishment of factories; with the first factory came the first company store. As the factories in the early period of their history were located remote from business centers, the store in connection with the factory was a prime necessity in order to supply the workmen with the essentials for their living. These stores at first were very crude affairs; the stock comprised only the very plainest articles of clothing and food, rough cotton goods, linseys, flannels, pork, flour, coffee, brown sugar, molasses, etc. The manufacturers soon discovered that with this forced trade and the

View of Glasstown plant office. Looking toward what is now Buck and Mulberry Sts. Taken about 1900. *Virgil Johnson photo.*

absence of competition very handsome profits were to be realized. The stock was very greatly enlarged and goods of better quality and higher prices added to tempt the wives and daughters of the employees to increase their buying. The passbook was very handy and convenient, so that at the end of the fire, when settling day came, the workman in many cases found that not only were his wages consumed, but that he was actually in debt to the store.

This condition of affairs tended largely to the benefits of the employers; a man in debt at the store was a man in bonds; he might fume and struggle, but he was firmly held; he could get no money, but must go on working

Photograph in front of Whitall-Tatum Company Store. From left to right: Miss Shaw, Edw. Dawnes, Ike Tompkins, Shropshire, Sam Fox, Ewan Cossaboon, Cora Sackwell Ritchie, Harry Thomas.

and find his wages consumed in his store account and in extinguishing a part of his indebtedness. What could he do? If he refused to work longer and left to seek employment elsewhere, he found that he was on the black list and could not get work in any other factory. If men wanted to get hold of a little money, they would get some article charged in the book at the store and dispose of it somewhere at perhaps half its value.

Prior to 1864 there was no law imposing any check upon the manufacturer in his dealings with the men, except the common law, which enabled an employee to recover his wages in an action of debt. Every possible device was resorted to by the manufacturers to avoid, as much as possible, the payment of cash; the employee received his pay in house rent and in goods out of the store. His demand for cash at intervals was frowned upon and he was expected to take the major portion of his earnings from the company store. There was practically no limit to his credit there, and if he was in debt, the condition was not unsatisfactory to his employer. The manufacturers tried, by every means in their power, to limit the payment of cash for labor; shinplasters took the place of legitimate currency and were redeemed at a discount. During the Civil War, many of the firms issued brass tokens to take the place of pennies, which had become very scarce, and, as they accumulated, redeemed them, frequently at a heavy discount. A Glassboro firm had issued an enormous quantity of these brass pennies, as they were called, and a businessman in Clayton had taken in a sufficient number to fill a nail keg. These he took to a firm in Glassboro for redemption; the firm refused to redeem them except at 35 percent discount, which he refused to accept. The brass tokens were large and heavy, and, as the price of brass had greatly increased, owing to its scarcity, the Clayton merchant sold the tokens to a dealer in brass for a cent and a quarter each. The Glassboro firm made the next issue on sheet brass, barely thick enough to hold together.

There was a reason for the reluctance of the manufacturers to pay cash for wages; money was scarce and rates to borrowers high. The wares sold on long credit, and often extensions were asked for, which he was compelled to grant, so that the obtaining of cash to settle with the men was a very serious matter. The store furnished a very convenient method of avoiding the payment of large sums for wages. The manufacturer could get long credit for store goods, and, as the prices charged were largely in excess of those charged in neighboring stores, he made a very profitable transaction in forced trade and high prices, and avoided, to a great extent, the payment of heavy discounts in obtaining cash for settlements. It is not difficult to appreciate the reluctance with which the manufacturer yielded to cash payments, or to understand why he should array himself against the efforts that were made by the men to secure legislation to enforce the payments of wages at fixed periods in cash. It is human nature to hold on to a good thing as long as possible.

The introduction of a bill in 1864, by Senator Providence Ludlam, of Cumberland County, was the beginning of a contest which has culminated in present laws prohibiting the forced trade and securing the payment of wages in cash, and this at periods definitely fixed. It is a rather remarkable fact that nearly all the legislation effected in this direction was the work of Cumberland County Representatives. The bill introduced by Senator Ludlam was finally passed after being amended so as to materially affect its efficiency.

Finally in 1899, the *Cash Bill,* which had the approval of the labor unions, was introduced by Senator Edward C. Stokes of Millville. Senator Stokes, who later became Governor, worked hard for the bill and insisted that before the bill was enacted into law it must have the en-

Shetterville (South Millville) plant of Whitall-Tatum Co. View shows south end of factory yard; iron stacks are for #11 & #12 day tanks. Brick stacks are #9 & #10 pot furnaces. *Virgil Johnson photo.*

Looking west from the gate of the Shetterville (South Millville) plant of Whitall-Tatum Co. Exact date unknown. Probably 1915–20. *Virgil Johnson photo.*

dorsement of the several locals of the Glass Bottle Blowers' Association. It finally passed the Legislature and became a law, and at the next conference in Atlantic City in 1900 the union representatives met with the manufacturers' representatives and in a "quiet and gentlemanly conference" discussed and agreed to restoration of the full wage scale, known as "the list"; modification of the rule governing the taking of apprentices; the abolition of company stores, and the release of requirement to live in houses owned by the company. It may have been the most important conference in the history of the trade.

The *Stokes Law of 1899* was thorough and complete in its provisions for insuring the payment of wages in cash. If it failed in any degree to secure the results aimed at in its enactment, if impositions upon workmen were not prevented, the fault was not in the law and must be looked for either in the failure of those for whose benefit it was enacted to invoke its protection, or the neglect of the officers whose duty it was to enforce it.

Company stores, however, did not go out of business. There were five in Millville quite a few years after 1900, but the workers were not obliged to deal out of the store unless they desired, although, sometimes if the employee thought enough of his job, it might prove helpful if the worker dealt at the company store. In Millville, Whitall-Tatum operated stores at Glasstown and Shetterville. Wheaton had one on North High Street, the Millville Bottle Works on Third Street, and the Millville Manufacturing Company at Green and Columbia Avenue.

CHAPTER 5

# Whitall-Tatum Company, Armstrong Cork Company and Kerr Glass

THE Whitall and the Tatum families trading under the names of Whitall and Brother, Whitall Brother and Company, and Whitall-Tatum Company began operations in 1834. Frank Shetter, from Baltimore erected the glass works on the present site of the Kerr Glass Manufacturing Company's plant in 1832.

Shetter failed in 1834 and Lewis Mulford with William Coffin and Andrew G. Hay of Winslow, New Jersey took over the operation and carried on under the management of Mulford. The firm name was Coffin, Mulford and Company. Window glass was made for one season under this management and after that they made green bottles and vials.

In 1854 Mulford decided that the glass business didn't pay and decided to make the Whitalls buy it. The Whitalls didn't want it, but when Mulford bought up all

Whitall-Tatum Co. glass works at Shetterville (South Millville) ca. 1900. *Chas. B. Pennington photo.*

South Millville plant of Whitall-Tatum Co. looking east from the Maurice River. Middle two stacks show pot furnace locations for large container manufacture. *Virgil Johnson photo.*

the wood in the vicinity, the Whitalls were unable to obtain any for use as fuel. They began to buy wood in Virginia and to transport it to Millville by boat but this was costly and inconvenient and they sometimes ran out of it, and so they were finally forced to buy the South Millville Plant. After the Whitalls bought the plant they turned it into a flint glass bottle factory.

In 1857 I. F. Whitall withdrew from the firm and for the first time the name of Whitall and Tatum appeared as the firm name. The new name was Whitall-Tatum & Company.

When C. A. Tatum entered the firm in 1875 he was with them for one year and then took charge of the New York office in 1876. It was in 1875 that the New York office located at 46 Barclay Street. The office was occupied until 1928 when the location was changed to 225 Varick Street. Prior to 1875 the firm had offices at 96 Beekman Street (1852–68) and 7 College Place (1868–1875).

On January 2, 1901 Whitall-Tatum Company was incorporated in New Jersey, the old firm being dissolved by mutual consent. All its assets were turned over to the new company which was to assume all of the liabilities of the old firm.

The first officers of the newly incorporated company were:

Charles A. Tatum, President
J. Whitall Nicholson, Vice President
John M. Whitall, Treasurer
Albert H. Tatum, Secretary

During the history of the company there have been relatively few managers considering the number of years it has been in existence.

The managers from 1836 to 1938 were:

William Scattergood, 1836–1845
I. F. Whitall, 1845–1848
Edward Tatum, 1848–1857
Henry Laurence, 1857–1865
Robert Pearsoll Smith, 1865–1869
John Mickle, 1869–1892
George S. Bacon (Glasstown), 1892–1897
Bond V. Thomas (So. Millville), 1892–1897
George S. Bacon (Both Plants), 1897–1939

George S. Bacon upon his retirement in 1939 was succeeded by his son J. Laurence Bacon. When the latter was assigned to Armstrong Cork Company in Lancaster, Pennsylvania as consultant to the Glass and Closure Division, W. O. Gassner was transferred from Lancaster, Pennsylvania to Millville as plant manager. Later W. W. Pedrick III succeeded Mr. Gassner and when Mr. Pedrick was assigned to Lancaster, Pennsylvania Roger F. Scott, formerly plant chemist, was promoted to position of plant manager, a position he still holds.

Whitall-Tatum Company entered the glass business before there were any mechanical appliances for glass making. The desired quantity of molten glass was gathered from the furnace on the end of a hollow rod, called a blow pipe, guided by the eye of the workman. The glass was then rolled back and forth on a flat metal plate or on a stone to partially shape the glass for the making of the bottle. This was called marveling. While this was being done, air was blown thru the pipe into the glass, making it ready for blowing the bottle. The blowing was accomplished by placing the glass in a mold and blowing it to the size determined by the mold. The molds in use were clay, cylindrical in shape and formed the body only. Now the partly finished bottle, but with the neck still unfinished and adhering to the blow pipe, was withdrawn from the mold and the end of a heated rod, or punty, was stuck to the bottom of the bottle. The blow pipe was then detached and the neck portion of the unfinished bottle was held in the furnace thru the ring hole and reheated to become plastic again. The excess glass was then sheared off by hand and hand tools were used to shape the neck and the flare of the mouth. The scar on the bottom caused by the punty was ground and polished out in the case of expensive bottles. Later a lip was finished on the neck by laying on a strip or ring of hot glass and smoothing it with a shaping tool. About 1850 a "snap" was used to hold the bottle while the neck, mouth and lip were being finished. This eliminated the scar on the bottom.

Up to 1865 everything was finished at the main furnace. All articles had to be reheated to lay on the ring. This was very slow work. Sixty dozen of one ounce vials could be made in ten hours, 55 dozen of two ounce and 48 dozen of four ounce. At this time every one worked alone and did all of the work at the furnace, both blowing and finishing. In 1865 the "glory hole" instead of the furnace began to be used to reheat the neck and

HAND GLASS BLOWING SHOP AT MILLVILLE STREETARAMA

This is the display that stole the show during the three day celebration of public improvements in Millville, New Jersey. Visitors saw Lorenzo Lober and Lewis Hund take gobs of molten glass from GBBA's portable furnace (1); gob was then "marveled" into approximate shape of container on flat metal surface (2); next step was to insert marveled gob into mold (3) operated by "mold boy" Romulus; from there "snap-up man" George Baumbach "sheared off" rough edges and re-heated mouth of container in "glory-hole" (4); final step was "gaffing" performed here by Wilson Pike (5) in which container mouth was shaped. In actual hand ware manufacture the containers were then removed for "tempering" a slow cooling-off process without which the container would be too fragile for handling. *Photo courtesy Glass Bottle Blowers' Association, AFL–CIO.*

South Millville workers ca. 1915. Unidentified.

blowers worked in shops of three men, two to blow and one to finish, the men alternating. The "glory hole" was fired with coal. At first men from four or five shops used one "glory hole." Later the glory holes were made smaller and were use by only one shop. Sometimes two snapping up boys were necessary for each shop. About 1891 the oil fire glory hole was installed and then only one boy was needed.

The next advance from hand blowing was the making of a complete bottle in a semi-automatic machine with hand gatherers, for both narrow and wide mouth ware.

The next step was getting the glass direct from the furnace thru feeders to automatic machines, which is now the universal method of making glass containers.

The earliest furnaces were built in the form of a horse shoe containing seven open pots arranged around the curved front and down either flat side. The firing was done from the back or flat end. The pots contained about 1200 pounds of glass each when molten. Oak and pine wood with rosin, with forced draft, were used as fuel to obtain the intense heat necessary to fuse the batch. Later cord wood was dipped in gas tar, and still later hard coal was used, still with forced draft. This type of furnace was used for colored glasses, mostly green.

For flint glass, the covered pot was used. While the firing of the furnace for green glass was done on the level of the pots it was necessary to fire from the bottom for flint glass. The fire came up into the furnace from underground caves. It was necessary to fire underground when making flint glass, otherwise the dust from the soft coal being used gave an amber tint to the glass.

In the late fifties the furnaces were rebuilt of an entirely different design, being circular in form and having a deep eye and a tall stack. With the tall stack, natural draft was used. In 1870 soft coal was substituted for hard coal.

The first covered pot furnaces had eight large pots and one small one known as a monkeypot. Later this was increased to ten pots. The average time to melt a pot of glass was about twenty-seven hours and the pots could be worked only every other day.

In 1893 a day tank was built at Glasstown for making opal glass. This tank was fired with oil. Later this tank was used for other colors and finally for flint. With the

Several early glassworkers pose for a photograph. From left to right: Frank Reed, Josh Owen, Chas. Crickler; Dave Lutes, Bill Bailey, Clarence Wescoatt, J. Webb, Neil Shaw; Jessie McHenry, J. Stanger, Bill Carmelia, Restore Daughtry, Sam Stanger, John Hand. *Virgil Johnson photo.*

day tank they melted at night and worked in the day time. For firing the day tank, producer gas was used. Here the fire played over the glass so it had to be a clean fire.

In 1893 a continuous tank was built at Glasstown with a daily capacity of from 20 to 25 tons. This was worked in three eight hour shifts. In 1893 crude oil was used as a fuel in all tanks. A few years later producer gas was used exclusively.

In 1897 a continuous tank was built at South Millville to make flint glass. Whitall-Tatum Company was the first company to make flint glass in both the day tank and the continuous tank, while still melting glass in pots.

The early furnaces all had names, given for various reasons and purposes, recording happenings in or around

Early view of Whitall-Tatum Glasstown Plant.

each furnace. These early furnaces had such names as Union House, Hen's Nest, Owl's Nest, York House, and Pigeon's Nest.

In the early days, the ingredients for the batch were taken just as they came. Later Richard Atwater maintained a so-called chemical laboratory from 1865–1890. This was the first laboratory in the glass industry for analytical work and all formulas for batch mixes were made according to analysis. Since 1890 the laboratory has become firmly established as a control in the manufacture of glass. No up-to-date glass plant could operate properly without a laboratory.

Most of the early molds were clay, although there were a few of iron. With flint glass some wooden molds were used.

In 1839, Thomas Campbell was employed by the company to make molds. In 1840 he made the first metal mold of brass entirely with hand tools. Iron molds followed as did crude machinery for making them. Around 1862 a mold making department was built at South Millville. There were separate departments at the two factories until 1889 when all work was moved to South Millville. In 1900 began the development of mechanical means for mold making. This has developed until the company has now one of the most complete and up-to-date mold making departments in the industry.

Both hollow ware (table ware and bottles) and window glass were made under the Lee management. In 1818 Gideon Scull converted the factory to the production of hollow ware entirely. In 1866 or 1867 the company went into the manufacture of general laboratory glassware, and wooden mold work. The lamp room was opened and South Millville began to grow. They began to make lettered plate ware in 1868 and shop furniture ware in 1871. Vials and lamp made ware followed in 1873. They began to make druggists sundries in 1876, and chemical ware in 1878. In 1878 they made perfume bottles, tooth powder bottles, glass pomade bottles, opal gallipots, engraved and cut glass bottles. They also made shelf bottles with labels attached for druggists. The company has always made patent medicine ware. They also made carboys in 3, 5, 10, 12, and 14 gallon sizes. Since the development of the semi-automatic and full automatic machine the company has made commercial bottles in flint, light green, emerald green, and amber. Before re-

View of the Glasstown plant of Whitall-Tatum Co. Taken in 1932. *Photo courtesy Armstrong Cork Company Archives.*

peal most of the business was prescription, pharmaceutical and a general line of commercial bottles. Since repeal 50% of the business has been beer and liquor bottles.

In 1899 A. S. Granger, a mechanical engineer, was employed to design means for manufacturing machine made bottles of all types. In 1900 an experimental plant was established. This work was done at South Millville under the direction of Benj. T. Headley and Mervin C. Bard. The experimental department was fully equipped with a glass furnace, tempering ovens and machine shop with a designing mechanical engineer and an expert mold maker in charge. Soon a machine for wide mouth bottles and jars was developed, to be followed by mechanical shears for cutting the glass from the source of supply. The entire operation of the machine depended on the gatherer who tripped a lever, setting the shear mechanism in operation, which in turn set the entire machine in operation. Soon mechanical means for opening and closing the molds were incorporated and later an automatic take-out was added. With a successful feeder the operation became entirely automatic. In 1911 a narrow mouth machine was designed and developed to commercial use in 1912.

Pressed ware such as lids, stoppers, and ointment pots was made on a side lever press. This method was slow and was a hand operation. In the late nineties a rotary press was designed and built. This had a round table mounting four molds. The press head was operated by hand by a side lever. In 1899 plans were made for a rotary press operated by air cylinders, making the machine semi-automatic. Mechanical cutoffs were added as soon as they were perfected. In 1921 engineers designed a rotary press for insulators, which would press the article and form the inside thread automatically. The molds were opened and closed by hand as was the transfer of the thread former from the screw out mechanism to the pressing mechanism. In 1926 engineers began work on a continuous rotary press. This was put into operation in 1928 successfully making insulators. The pressing and blowing machine was developed in 1904.

Annealing, which is the method of relieving strains in glass was first done in ovens fired with wood. The glass was put in the oven, sealed, and allowed to cool as the wood burned out. This process took three days. These ovens were used for many years. The next advance in annealing was the continuous lehr, fired first with wood, then coal, then producer gas and oil. These first continuous lehrs were open fired from the bottom and the fire came over the top. This took two days. Muffle lehrs were next developed, the heat being confined to chambers. This was much cleaner and kept the dirt off the ware. Oil was used as well as producer gas. The annealing was reduced to three hours. At first the bottles were transferred from the shop or machine to the lehr by carrying in boys. With the development of the unit lehr, the ware began to be automatically transferred to the lehr by conveyor belt and placed in the lehr automatically by a stacker. This latest development made the operation entirely automatic. The heat control of both furnaces and annealing ovens and lehrs was first determined by the skill of the shearer who judged the temperature by his eye and experience. Temperature control means were gradually developed until now all melting and annealing is done under control of highly sensitive pyrometers.

The ware was first packed in wooden boxes or barrels with hay or straw. Later newspaper replaced the hay and

Whitall-Tatum schooner *Caroline*, used prior to railroads for cargoes of finished ware to New York and Boston. Lost at sea in 1878. *From an early engraving.*

straw. The use of wooden boxes and barrels was gradually replaced by corrugated boxes and today most bottles are packed in this way.

The first shipping was done by water, both raw materials and finished goods being moved in this way. The company owned two sloops, two schooners, and a steamboat. The Sloop Ann went to Philadelphia, the Sloop Franklin to Philadelphia and Baltimore, the Schooner Caroline went to New York as did the Schooner Mary and the Steamboat Millville. Early shipping was done by teams. Even after the railroad came to Millville in 1863, the ware was carted to the depot by team. Incoming freight was handled in the same way. In 1881 there were

The engine *Eagle*, the first locomotive on the Millville & Glassboro Railroad (1859), later the West Jersey Railroad. The engine was brought to Millville on a schooner. *Granville Thomas photo.*

twelve teams at Glasstown and 18 at South Millville. Later the railroads were brought into the two plants and incoming and outgoing freight handled direct. The boats were gradually abandoned for the more modern way of handling, and the teams finally gave way to trucks, tractors and electric trucks.

As was the custom in the early days, the company operated company stores. The stores dated back to the days of Burgin and Pearsoll. The South Millville Store was built in 1878. In the early days everything was bought from the store. Everyone was paid off here, getting money as he asked for it. Every two weeks they ordered the money they wanted. At the end of the year a settlement was made and the rest of their money given to them. The stores were finally closed in 1916 when it was no longer possible to deduct a man's bill from his wages.

In the early days the buildings were low factory buildings and warehouses, all of wood.

The company was justly proud of its personnel and the record of service attained by some of its employees. In 1916 the company had 32 men on its payroll who had served 44 years or more with them, entering the employ at the average of 11 years. Samuel Berry completed 76 years of service in September 1938.

In the early days boys went to work at the age of ten. As they showed ability, they advanced to care for the ware in the tempering ovens or the foot bench to gather the glass for the blowers. At sixteen they were eligible for apprenticeship for a five year term at half wages. It was long the company policy to develop its own workmen and make promotions within the organization. For this reason, few men or boys voluntarily left the company's employ. This resulted in small turnover and a great many employees with long service records.

Whitall-Tatum Company was developed thru all stages of the glass industry, starting in the days when all ware was made and handled by hand. It had seen and taken part in the development of furnaces, improved glass, lehrs, machines and glass feeders until today as the Kerr Glass Co. it is one of the best equipped plants in the industry for the automatic manufacture of glass containers of all descriptions.

Up to 1938 when Armstrong purchased Whitall Tatum Company South Millville had over forty buildings and 428,172 square feet of floor area.

The present plant has sixty buildings with 939,000

square feet of floor area. They manufacture 400 different types of containers.

There are presently five furnaces and twenty-four machines at the plant. A machine may produce from 350 gross per day to 1300 gross per day, depending upon the item. The employment is approximately 1350 persons, including clerical and staff. Glass is shipped by truck and rail with most of the shipping done by truck.

There are forty-five acres of land within the fence. There are twelve warehouses for finished goods. Approximately 1200 pallet loads of containers are produced and stored per day (24 hours).

An agreement in principle for the purchase of Armstrong Cork Company's Glass Container Division by Kerr Glass as of April 1, 1969 was announced February 1, 1969 by the following press release:

> Lancaster, Pa. (February 1, 1969) Agreement in principle was reached here yesterday for Armstrong Cork Company to sell its Packaging Materials Operations to Kerr Glass Manufacturing Corporation of Los Angeles, California. It was jointly announced by James H. Binns, president of Armstrong, and William A. Kerr, president and chief executive, Kerr Glass. Closing of the transaction is scheduled for April 1 upon satisfactory accomplishment of the terms of a definitive agreement. No problems in closing are anticipated.
>
> The agreement in principle, subject to the approval of the boards of directors of the two firms, calls for Kerr Glass to acquire the Armstrong packaging materials plant facilities and related assets. Included are the Lancaster, Pa. and

Arlington, Texas closure and plastic container manufacturing plants; the Millville, New Jersey; Dunkirk, Indiana; and Waxahachie, Texas glass container manufacturing facilities; the Keyport, New Jersey closure and plastic vial factory; the Charlotte, North Carolina plant site, and the assets of the National Cork Company, exclusive of cash, receivables and its insulation contracting business.

With the addition of the Armstrong facilities, Kerr Glass, with a new total of 11 plants, would become a nation-wide manufacturer and distributor of glass and plastic containers and metal and plastic containers and metal plastic closures.

Mr. Binns said that the sale of Armstrong's packaging business represents a major policy decision which implements Armstrong's corporate strategy of concentrating its assets and resources in the markets it is best suited to serve. "While our packaging materials operations are successful—and our manufacturing facilities are among the most modern in the industry—our share of the packaging materials market is relatively minor, and it is obvious that a continued program of expansion would be required for Armstrong to gain a substantial position in this industry.

With expansion in packaging as one alternative, we have to consider opportunities for growth in a number of other markets of great interest to us. We have concluded that certain other areas of opportunity, particularly suited to our talents, capabilities and resources, offer greater potentialities for future investment."

As one example of Armstrong's strategy of concentrating its efforts in markets, it is best equipped to serve, Mr. Binns cited the company's strengthened position in the interiors field which resulted from the acquisition of E & B Carpet Mills, Inc. and Thomasville Furniture Industries, Inc. He added, "With other investment opportunities having priority in our future expansion plans, we would be in effect restrict-

Aerial view of the present manufacturing facilities of the Kerr Glass Manufacturing Company located on the original site of James Lee's first glassworks. *Photo courtesy Kerr Glass Mfg. Co.*

ing the futures of the loyal and competent people who are associated with our Packaging Materials Operations. We are delighted that the move to Kerr will solve this problem by providing our people with expanding opportunities in their field of primary interest."

Mr. Kerr said that his company recognizes, just as Armstrong does, that continuous expansion in packaging is fundamental to sustained success. "Because packaging is our field of specialty, we are determined to expand and become one of the major factors in this dynamic growing industry," he said. "We have great respect for the Armstrong packaging organization and we are delighted to welcome them to our company. We are confident that these highly qualified specialists will find enlarged opportunities in the expanded Kerr Glass Manufacturing Corporation."

Mr. Kerr said that when the transaction is completed, Mr. W. W. Pedrick III, formerly of Millville and now an Armstrong director and vice-president and general manager of its Packaging Materials Operations, will join the Kerr organization as a director and a senior vice president. He said that the former Armstrong facilities would be operated as an autonomous unit under Mr. Pedrick's direction from eastern headquarters which will be established in new facilities in Lancaster, Pa.

Kerr Glass Manufacturing Corporation, began operation in 1903. It employs about 1700 men and women and has manufacturing facilities in Santa Ana, California; Huntington, West Virginia; Chicago and Plainfield, Illinois; and Sand Springs, Oklahoma.

Armstrong Cork Company was founded in 1860. It is a diversified manufacturer of building products, resilient flooring, furniture, carpet, ceilings systems, packaging materials, insulation and industrial specialties.

Thus Armstrong Cork Company's glass container operations have joined those of the illustrious Whitall-Tatum Company in Valhalla. We will always recall these names for both have been symbols of the best in men's souls. In parting with friends Millville also welcomes Kerr Glass Manufacturing Corporation. Its predecessors built on firm ground, a heritage now placed in the hands of the new company.

CHAPTER 6

# A Retired Glass Container Executive Journeys Down Memory Lane

WILLIAM PEDRICK II, now 86, a resident of Main Rd., a short distance north of the Millville city line, is one of the few men about the Millville area, who possesses the kind of memory that can take him back to 1901 when he finished Millville High School and got a job as office boy at the Shetterville plant or lower works of the Whitall-Tatum Company.

George S. Bacon was superintendent of the two plants, upper and lower, or Glasstown and Shetterville plants of the Whitall-Tatum Company. Pedrick retired in 1948 as production manager of the South Millville plant.

"When I began work for Whitall-Tatum Company, there were 2400 employed at South Millville and 1400 at the upper plant," said Pedrick.

"Today, with automatic machines, approximately

1400 employees are in a single plant, at South Millville, more than five or six times the number of bottles are produced than there were when I worked there. It is amazing the way those machines turn out ware. Sam Berry was boy boss and Harry Woodruff was boss of the blowers. Lew Corson, father of Millville's late Mayor, Benjamin Corson, was paymaster. Whitall-Tatum Company had a New York office and had 42 salesmen on the road.

"The company used to freight out glass bottles by boat as well as by freight car, but that was a little before my time, and some years before I went to work, the furnaces were fired by wood.

"Then came coal, then oil, gas and now there is an all-electric factory at the Wheaton Glass Company's plant that started this past fall."

Mr. Pedrick may be one of the few persons, living today, who can name the people on a photograph made in the 1905 era of the South Millville office force. There are several pictures of that group around town but there seem to be few who can name them all. Pedrick names them as follows:

"Rhetta Corson, Mamie Stein, Emma Godfrey, Nellie Ward, Bessie Evans, Cecelia Campbell, Ada Vineyard, Mae Rieck, Bertha Bard, Ralph Hoffman, and Charles Bartlett.

"Alex Querns, Edward Downes, Mort White, Lewis R. Corson, Sam Berry, Sam Fox, Sam Niaish, William Calkins, "Jerry" Corson, Albert Ritchie, Frank Pierce, Harry Heintz.

The South Millville office force, Whitall-Tatum Co., ca. 1908. *Virgil Johnson photo.*

"Hammill Thompson, Jack Kauffman, Ralph Getsinger, William Loper, George Peacock, Walter Stites, Frank Conover, Ben Headley, Dubray Masters, and Howard Bomhoff.

"Josh Corson, Jack McClure, William Nicholson, assistant manager and Boyd Henderson."

Mr. Pedrick worked at Shetterville when those expert glassblowers, John Fath Sr., Amy and Tony Stanger, brothers; Marcus Kuntz, John Rhulander, Ralph Barber and others, many of them coming from Alsace-Lorraine, Europe, worked in the South Millville plant.

He tells of how Jack Horton, a butter and egg man, was hired by George S. Bacon and later became a great friend of Mr. Bacon who made him plant manager at Glasstown.

It was the era when most every boy's desire was to become a glass bottle blower. They were called "those millionaire glassblowers in Millville" in other South Jersey communities. They were able to earn from $5 to $15 a day when the average pay in other forms of employment would be from $7 to $10 and $12 a week.

When I was a freshman in Millville High School, Bill Pedrick was an upper classman and a very good high school baseball pitcher. That was just before he began working at the South Millville plant of Whitall-Tatum Company.

He tells me that Sam Berry started working at the Glasstown plant when he was eight. He was 88 when he died and death came a few years after he retired. During his career, he learned the trade of glass bottle blowing. He later became a foreman in charge of "boys" at the South Millville plant.

"Boys" employed in glass plants around the turn of the century were sometimes mere "kids" eight or nine years old and while it was illegal to hire boys under 12, it was done. When a state inspector would visit the plant, the kids were told to hide. Many times, it was said, the foreman would hide them under a barrel while the inspection was being made.

Adults who had four or five boys in their family were more likely to obtain work at a glass plant in those days than a man who had no boys. That is the way "Col." Evan Kimble, who became a millionaire in the glass business, got his start. He was one of five boys in the Kimble family. The Kimbles lived in a company-owned house in what was called Bucktown, just west of S. Second St. and north of the railroad spur that enters the South Millville plant yard and all the boys worked in the factories.

Jess Cossaboon, a retired policeman, now dead, used to tell of the influence Whitall-Tatum Company had. It even extended into the schools. Cossaboon was a pupil in the South Millville School, of which Silas Smith was principal and he figured it was good business to cooperate with Whitall-Tatum Company. He did. If Sam Berry was short of boys, he would send word to Principal Smith that he needed four or five boys and Smith would call out Jess Cossaboon, Jimmy Jones, Harry Brown and Billy Morrison. "You're needed over at the Shetterville factory. You are excused. Report to Mr. Berry immediately." Jess said, "We were glad to get out of school and we hustled over." Everybody or most everybody wanted to be a glassblower those days.

Pedrick tells of Bond Thomas who was, one time, manager of the South Millville plant and how he quit to accept a better job with Tiffany of New York City. He tells of Amy Stanger, an expert glassblower, making blue bowls for Tiffany. He made one out of cobalt blue, off-hand ware, that might have weighed 12 pounds.

Pedrick may have the only cobalt blue bowl around. It was a company sample and was given him by George S. Bacon.

Pedrick also remembers Harry Bard and Park Thompson having designed changes for the bottle machines, called the automatic cut-off that saved the company the cost of a man and the company obtained patents on their inventions. There was court litigation but the ruling was that the men were employees of the Whitall-Tatum Company and therefore the company was entitled to the patents. Thompson and Bard received a small return for the invention which could have made them rich. Similar ideas by the company chemist George Barton, saved the company many thousands of dollars but did not benefit Barton to any great extent.

It is interesting to listen to Pedrick tell stories of the days of the company store, company home, the boy-workers and when the glassblower was a kind of king around South Jersey. He earned more money than any other worker.

CHAPTER 7

# People in the Early Days

## George S. Bacon

GEORGE S. BACON may have had more influence in Millville from 1897 to 1938 than any other single citizen.

He was, in that period, general superintendent of the two large *Whitall-Tatum Company* glass plants in Millville, the Upper Works, located on the area of the present American Legion Home, and the Shetterville, or lower plant, on S. Second St. Nearly 3,000 workers were employed in the factories.

In the early years of these factories, only men were employed and only men could vote.

In the Third Ward was located the *Millville Manufacturing Company's* mill, bleachery and iron foundry, during a portion of this period. These industries were owned by the *R. D. Wood Co.* and the two companies were involved in a continuous struggle for the control of the city with *Whitall-Tatum Company* generally having the edge.

George S. Bacon was a large man, six feet, two or three inches tall. A short black beard and his erect posture helped to give him a dignified appearance. A Quaker, born in Bacon's Neck, Cumberland County in August 1864, he was graduated from West Town Quaker School and obtained employment in the Philadelphia office of *Whitall-Tatum Company* in 1888 and was sent to Millville by the company later in November 1891. He was placed in charge of the Glasstown plant in 1892 and in 1897 he was made superintendent of both plants. He was elected vice-president in charge of production in 1925. He retired in 1938 and he died March 9, 1944.

He married Miss Rebecca Mulford and from that union, there were three daughters and a son.

The Mulfords were, at the time, one of the city's prominent families. Mr. and Mrs. Bacon set up housekeeping in a residence at 116 N. Third St. They later moved in a large residence on Columbia Ave., still standing near the Vine St. Bridge.

Two daughters and a son were married.

Miss Margaret Bacon married Charles Ford Adams, now deceased. There is a married daughter, four grandchildren and one great grandchild.

Lawrence Bacon married Miss Dorothy James, of Philadelphia, Pa. and is living in Lancaster, Pa. He is the father of two sons, James, of Millville and George, of Landover, Md. There are three grandchildren.

George Bacon. An early portrait. *Photo courtesy Dr. Elizabeth Bacon.*

Miss Caroline Bacon married Chandler Burpee, now deceased and their home is in Hampton, N. H. There are three sons, all married and 10 grandchildren.

Dr. Elizabeth Bacon resides in Millville and is a member of the faculty of the Millville Senior High School.

The Bacons contributed much to the progress of Millville. Mrs. Bacon gained the reputation of doing a tremendous amount of charity work around town and it was principally because of her interest among the unfortunate families here that the Board of Education named the school erected on S. Third St., on a site provided by *Whitall-Tatum Company* through the influence of Mr. Bacon, the Rebecca Mulford Bacon elementary school.

When George S. Bacon retired after the *Armstrong Cork Company*, of Lancaster, Pa., acquired the two *Whitall-Tatum* Company plants in 1938, his son, Lawrence Bacon was placed in charge by the new owners. In 1944, he was moved to the Lancaster offices and William Gassner, of the Lancaster headquarters, was sent to Millville in May 1944. William Pedrick III, a graduate of Vineland High School and the son of Mr. and Mrs. William Pedrick II, Main road, became plant manager on July 1, 1949.

His enthusiasm and determination earned promotion for him in February 1957 when he was placed at the head of the company's closure division and moved to Lancaster. He is now a company vice president.

Roger Scott, present plant manager, succeeded Ped-

George Bacon, General Manager, surveying the damage after the great fire at the Whitall-Tatum Co. Upper Works, ca. 1920. *Virgil Johnson photo.*

rick. Originally, a farm boy from Greene County, Pennsylvania, he was graduated as a glass technician from Penn State. After he obtained employment at the Lancaster offices of *Armstrong Cork Company,* he was sent to Millville to work under George E. Barton, head chemist at the South Millville plant for many years.

So, when Pedrick moved to Lancaster, Scott eased into the plant managership. In the years he has been in charge, there have been many and varied changes, including a new office building, more and better warehouses and more employees added.

Mr. Scott married Miss Jane Burroughs, his high school sweetheart. They are the parents of three sons, Daniel, at home; Donald, a graduate of the United States

Naval Academy, employed in Los Angeles Cal., and Thomas, a teacher in Massachusetts.

## James Lee

James Lee a son of Francis Lee, who came from Belfast, Ireland, to Pennsylvania, was born in 1771. He was an active, energetic man, having the impulsive character attributed to the Irish, but appeared to be of a restless disposition and was always abandoning one newly established enterprise for another. About the year 1799, he began to manufacture glass at Port Elizabeth, in a factory which he called the Eagle Glass Works. A few years later, about the year 1806, he erected the first glass factory in Millville, and shortly afterward sold his factory there and engaged in the same business at Kensington, near Philadelphia.

In 1814, Lee moved to Bridgeton and joined with Ebenezer Seeley and Smith Bowen in erecting the dam on the Cohansey Creek, known as Tumbling Dam. It was the design of Lee and Seeley to erect a paper-mill on the east side of the creek, the power to be supplied by water brought down in a raceway from Tumbling Dam pond to a site near the foot of North Street. Paper at that time was very high priced, owing to the war with Great Britain. A raceway was dug near the east bank of the creek, but Lee and Seeley ran out of funds and the project failed. They, however, built a saw-mill on the site of the proposed paper-mill, which they carried on for a time.

An advertisement appearing in a Bridgeton newspaper in 1816, tells of a store run by Lee at Laurel Hill, probably near his mill at the foot of North street.

NEW STORE
AT
LAUREL HILL

JAMES LEE has just opened store at this place, where he intends keeping for sale a general assortment of Dry Goods, Queen's and Earthen Ware, Groceries, Hard Ware, Paints, Oil, Medicines, etc. Which he will sell on liberal terms for cash, or on a reasonable credit, or barter for any kind of country produce.

WANTED TO EMPLOY from 10 to 20 good hands to dig a race way from the new mill dam on Cohansey to Laurel Hill.

Likewise, wanted four apprentices to the glass manufacturing in Philadelphia, at James Lee & Son's Works.

J. Lee.

Laurel Hill, Bridgetown,
May 23d, 1816.

In 1817, Lee removed with his family to Cincinnati, and from thence after a short stay at the latter place to Maysville, Kentucky. He died in New Orleans in 1824.

### Frank Shetter

Frank Shetter purchased a tract of land, described in 1832 and a quarter mile south of Millville and built a glass factory on the site. He and some friends came from Baltimore after learning that there was woodland and a navigable river running from Delaware Bay to Millville.

He built company-owned houses around the factory and many of them were still standing during World War I.

There weren't many houses south of Smith Street and much of the area between what was then known as Millville and the site on which Shetter built his hamlet which came to be known as Shetterville, was vacant.

As the years passed, homes were built north from Shetterville and south from Millville and by 1885 the intervening space between Shetterville and Millville joined and it became one community but it never lost the name of Shetterville although, today, only the older residents call the section south of the "First Hollow" Shetterville.

At the time, the only glass factory in Millville was the one at Glasstown, now the site of the American Legion Home and eventually the Whitall-Tatum Company purchased the South Millville plant after Shetter failed in 1884.

## William S. Breeden

The name of William Sheppard Breeden, born May 30, 1872, must be included among any list of Millville men who were active in the glass manufacturing business. He was one of five boys and one girl, a sister, Mary. He was the son of the late John Breeden and Miss

Beulah Hann, whose parents operated the *Hann Hotel* on the northwest corner of High and Main streets.

Will Breeden was educated in Millville schools and he worked at the *Whitall-Tatum Company* factories when he was a young man. The family moved to Millville when he was a youngster. While employed in the glass factory, he attended night school at Temple University, in Philadelphia. He also worked at the *Wheaton* factory for a time.

He married Miss Flora Risley, Mays Landing in 1895. Mr. Breeden started a small lamproom here just prior to his marriage. It was prior to the turn of the century when he, Angelo Giuffra and others joined in the establishment of the *Capital Glass Company* in the Vineland area. He sold his interest in that venture and went to Bradford, Pa. to take advantage of the natural gas. He returned to Millville. He worked with Paul Frederick and the *Mulford Company* for a time, then, in 1914, he started the *Eastern Glass Company* in the old *Dix Wrapper* factory in the rear of Buck Street, just north of Pine Street, employing an average of 50 workers. It was his most successful operation. The corporation included members of his family. A son, John was secretary. He is still living at 220 W. Main Street. Fancy glassware was in great demand and he operated until the depression, when his father closed the factory. William Breeden was the father of eight children. Seven are living.

During his residence in Millville he was active in civic affairs and at one time served on the Board of Education. He died in 1941.

## David Irvin DuBois

David Irvin DuBois, 79 years of age when he died, February 2, 1966, was an experienced glass man born in Dutch Neck, just outside of Bridgeton, was graduated from Bridgeton High School and Stroudsburg College and went to work in *Shaw's Mould Shop* in Bridgeton. He accepted a job with the *Dominion Glass Company* in Montreal but was lured back to South Jersey by a tempting offer from the *Minotola Glass Company*. Then he came with *Whitall-Tatum Company* who sent him to Stroudsburg.

In the meantime, he married Miss Abbie Patton, of Bridgeton and from the union, there were five children, Mrs. Aleta Lore, Trenton; Earl, Millville; Mrs. Doris Kane, Millville; Mrs. Edna May Ewan, Millville and William. There were 15 grandchildren living.

When the Stroudsburg glass works closed in 1917, he was recalled to the *Whitall-Tatum Company's* plant in Millville. He became chief engineer under George Bacon, superintendent of the company's two Millville plants. William Nicholson was assistant superintendent and George Barton was chief chemist.

When *Armstrong Cork* took over the two plants in 1938, DuBois was called to the Armstrong headquarters office in Lancaster, Pa., and asked what was needed and how much it would cost to get the Millville operations on the right track. In less than three days, DuBois provided the answer and the estimated cost was set at three million dollars. The program included the dismantling of the upper works and moving the equipment to South Millville.

There came more changes in 1941. George Bacon had retired. Lawrence Bacon was transferred to Lancaster and Harry McDonald was sent from Dunkirk to Millville to succeed Bacon. DuBois became assistant manager.

As Mr. DuBois recalls, all pot furnaces at South Millville were closed out in 1921–22 and they were gradually torn down. In 1927, No. 9, the second continuous tank at Shetterville, was built. No. 8, the first, was built around the turn of the century. They were called continuous tanks, but only two shifts worked on them for several years. The work week was expanded to six days with Sunday off.

Trouble with the glass Monday mornings may have been one of the reasons why the companies decided to operate continuously. That's the way it is today.

## Colonel Evan Ewan Kimble

Col. Evan Kimble, one time Millville resident, who was laid off when work became slack at the Shetterville plant of Whitall-Tatum Company in 1900, obtained a job in a glass factory in Gas City, Ind., left town, and never returned again as a resident. But he did come back to Cumberland County—came back, organized Kimble

Glass, at Vineland, and developed the experiment into one, if not the largest, manufacturer of chemical glassware in the country.

His career was a rags to riches story. He became a millionaire. He might have been a retired glassworker living on social security and his savings, had he remained in Millville.

"I was born in Tuckahoe," said Col. Kimble. "There were five of us boys and I guess it was when I was about five that we moved to Vineland. It may have been near 1880 when we moved to Millville. When Whitall-Tatum Company found out mother had five boys, we were all offered employment and a house in what was then called Bucktown which was a little settlement of company houses northwest of the company store, set back from Second St. Charlie Pepper lived there. We depended on Whitall-Tatum for everything—a home, wood, food, doctor and what little money we could get. I learned the lamp room trade there. I was just beginning to get along nicely when the slackening up of orders occurred and I was laid off."

As it turned out, that was the break that changed his life. At Gas City he became successful and became manager of the Sheldon-Foster Glass Company which, however, closed down. He moved to Chicago with the financial blessing of his former boss and started a lamproom known as the Kimble Glass Works which progressed encouragingly. But he experienced difficulty getting enough tube. That was his real reason in coming to Vineland. He acquired the Vineland Glass Tube Company and from that small beginning Kimble Glass Company came into being. The assets of the company were later acquired by Owens-Illinois and now, carries the name of Kimble Products Division of Owens-Illinois.

Mr. Kimble married Carrie, daughter of Mr. and Mrs. George Dougherty. He was a Methodist and a strong Republican. After serving in the New Jersey National Guard for several years, President Warren Harding made him a Colonel.

He was the kind of man Millville folks liked to say, "Col. Kimble learned the glass business in Millville." It is truly a story of "from rags to riches" because there was a time when the attic window in that tiny company house in Bucktown with a pane of glass broken out was stuffed with rags.

Col. Kimble died in 1957.

## Joseph Doughty Troth

Joseph Doughty Troth, one of Millville's most prominent residents, was a glassblower who, during his life, held the top office in his trade. He also became prominent in Millville politics and held important offices gained through his prominence in the Republican Party. He was born in Waterford and died in Millville in 1929 at age 72. His wife was the former Ida McLauglin.

Mr. and Mrs. Troth resided on the northwest corner of Fifth and Mulberry Streets. Their family included two daughters, Ethel and Hazel, and three sons, William, Leroy, and Sewell. Only Sewell is now living and he is confined to a bed in Ivy Manor, Bridgeton, with an arthritic ailment.

As a glassblower, a trade he learned at the Whitall-Tatum Co., he represented Local Seven as a delegate to the annual conventions of the Glass Bottle Blower's Association of the United States and Canada. He held offices in the Local and in 1891, was elected Vice President of the GBBA. In 1894 he was elected President of the GBBA at the National Convention in Montreal, Quebec. He was re-elected at the convention held in Atlantic City a year later. He was defeated for a third term by Dennis Hayes.

Mr. Troth was elected a member of City Council from the Second Ward for several terms and he also served as president of City Council and as County Republican Chairman. He was appointed postmaster of Millville and served a full term.

Mr. Troth was proud of the part he had in starting Edward Casper Stokes on his successful political career. It was the era of county political conventions and that year the Cumberland County Republican Convention was held in the Metropolitan Hall in Vineland. It was there that Mr. Troth nominated E. C. Stokes as the Republican candidate for the New Jersey State Assembly.

From there, Stokes went on to become State Senator and then Governor.

Mr. Troth was a Millville glassworker who had an imposing public career.

## Dr. T. C. Wheaton
## Frank H. Wheaton, Sr.
## Frank H. Wheaton, Jr.

One of South Jersey's largest and most successful glass manufacturing plants was started by a native of Tuckahoe, Theodore C. Wheaton. He grew up in a family of five boys. His success was the result of long years of struggle, hard work and determination, and as a boy, he never dreamed of owning a glass house. His father, Amos Wheaton, was a millright, a carpenter, who moved his family to Ocean View when Theodore was six.

The roots of the Wheaton family grow deep in the history of Cumberland and Cape May Counties. Joseph Wheaton, the great grandfather of Dr. T. C. Wheaton, was an officer in Captain James Willets' Company in the War for American Independence.

Theodore Corson Wheaton was born at Tuckahoe on August 24, 1852. His father, Amos, was a carpenter who gave up his trade in 1858 to manage a gristmill at Ocean View. From the latter place a boy could see the sails of the ships as they plied the waters along the Atlantic coast. At the age of 18 Theodore Wheaton shipped before the mast on a 400 ton coastal freighter. Because of seasickness, his career as a sailor was short. He took a job with the railroad, helping to lay the first tracks into Cape May County. Next he apprenticed himself to one Dr. Way, a pharmacist and physician at South Seaville. By the time

Dr. Theodore C. Wheaton. *The Foschi Studio, Vineland, N.J., courtesy Wheaton Industries.*

he was 21, Theodore Wheaton had saved a thousand dollars.

He enrolled at the Philadelphia College of Pharmacy and Science. Boarding with a pharmacist named Frank

Frank Hayes Wheaton, Sr. *The Foschi Studio, Vineland, N.J., courtesy Wheaton Industries.*

Hayes, Theodore earned his keep by working nights and weekends in the Hayes drug store. He graduated in 1876 and immediately enrolled at the Medical College of the University of Pennsylvania. While attending medical school he continued to live and work with Hayes. Some time during these busy years he managed to meet and court Bathsheba Brooks Lancaster.

In the spring of 1879 Theodore C. Wheaton, M.D.

returned to South Seaville to begin the practice of medicine. The following year he and Bathsheba Lancaster were married. In 1881 their first son was born. They named him Frank Hayes Wheaton.

Cape May County offered few opportunities for the ambitious young doctor and his city-bred wife. In 1883 Dr. Wheaton moved his family to Millville. Here in addition to continuing his career in medicine, Dr. Wheaton established a pharmacy and a store where general merchandise was sold. After a few years he sold the drug store and established another one further down town. Mrs. Wheaton bore three more children. Theodore C. Wheaton, Jr., Ada Bathsheba Wheaton and another daughter who died in infancy. Dr. Wheaton became involved in the civic life of the community serving, at different periods, on the City Council and the Board of Education.

He became interested in the activities of four local glassblowers who had erected a small furnace in the northern part of the town. One by one Dr. Wheaton bought out the interests of these men until in 1889, he owned the business. From this point on Dr. Wheaton gradually divested himself of his other interest and professions in order to devote all of his time to glass. To provide land for expanding his new venture, Dr. Wheaton purchased what amounted to half of Millville north of Broad Street. The southern part of the tract he sold as building lots, retaining the upper area for business expansion.

Next to his home, Dr. Wheaton established a "company" store where employees of the T. C. Wheaton Co. could buy their food, clothing and dry goods.

With Charles K. Landis, Dr. Wheaton pioneered Sea Isle City, New Jersey, a summer resort in his native Cape May County. The doctor and his descendents maintained summer homes in that place and it was there that Dr. Wheaton died on Labor Day in 1931.

Frank Hayes Wheaton, Jr. *The Foschi Studio, Vineland, N.J., courtesy Wheaton Industries.*

Frank H. Wheaton, Jr. is president of the Wheaton Glass operations, having taken over the responsibility two years ago. His father, Frank H. Wheaton, Sr., always ready to credit his son with being responsible for the amazing and dynamic growth of the Wheaton enterprises, has refused to claim any credit for their great expansion.

Frank, Jr. was graduated from Millville High School and took several short college courses while learning the glass business from the ground up. He spent several years in the factories, the offices and on the road to gain a thorough knowledge of marketing and manufacturing.

Presently he is in charge of the Wheaton industrial empire which is said by some to be the largest private industrial corporation in the United States. Born in Millville he is married, resides in Millville and is the father of three daughters and a son, all married.

Mr. Wheaton is also a director of the Atlantic City Electric Company, the South Jersey National Bank, Cumberland County College and the Millville Hospital.

CHAPTER 8

# Mr. R. Pearsoll Smith His Contribution to the Workingmen of Millville

R. PEARSOLL SMITH made a substantial contribution to the moral standards of Millville in 1881 when he supplied money for the erection of the Workingmen's Institute, now the old City Hall.

Mr. Smith was, at the time, associated with the Whitall-Tatum Company, owners and operators of two glass manufacturing plants in Millville. A rather detailed account of Mr. Smith's interest and activity is contained in chapter six of the Bureau of Statistics of Labor and Industries of New Jersey. The date of the issue is 1882.

A portion of the report states, "The liberality of a distinguished citizen of Millville has enabled the working men of that pleasant South Jersey town to establish an institute and erect a suitable clubhouse."

(The idea had originated in England where "Workingmen's Clubs" had been established under the "Workingmen's Club and Institute Union").

The chapter continues, "The considerations which induced Mr. R. Pearsoll Smith to assist in this good work are, as set forth in the following sketch of the Millville Workingmen's Institute, contributed by a citizen of that city:

"Situated in the sand barrens of South Jersey, Millville has few wealthy residents. It is a city of working people, possessing no men or women of leisure. In weekdays, during the day-time, its streets are almost as quiet as Sunday but, at night, they swarm with the operatives of the glass works, bleachery and iron foundry. A few years ago, like other manufacturing towns, Millville had its full quota of drinking saloons in which a large portion of its wealth and intelligence was wasted. Like other towns, too, it had its temperance agitators, who 10 years ago, determined to make renewed efforts to prohibit the sale of liquor. That, which characterized its work, was its success. Accomplished by the workingmen themselves, whose fairly large wages gave them the means with which to fight the liquor interests, it had the employers and clergy; while its permanence was assured by years of watchfulness and vigorous enforcement.

"The results of the work are now evident. Millville has a fine hotel, managed on strictly temperance principles. The church capacity is exceptionally large. Its building and loan associations are the most successful of any in the country and most of the skilled workmen own beautiful homes. A poor fund established some years ago had no call for its income. Its municipal government is without scandal and except an occasional liquor seller, it sends no one to the town jail. The license question has disappeared from city politics, the city council being unanimous against license while the police consider it their duty to hunt down illicit venders of liquor.

"Ten years of such results can be considered a fair settlement of the temperance question but, to thoughtful minds, the victory was not fully assured. The restless mass of artisans who were abroad upon the streets after their day's toil had been finished, sought some entertainment as a diversion from labor. If other forms of enjoyment were not provided, they would be certain to demand the attractions which may be found in the drinking saloon.

"The success of the women of the Christian Temperance Union in establishing a boys' club, built for the purpose by kind friends, suggested the practicability of a club for adult workingmen.

"A gentlemen of means, one of the manufacturers (Mr. R. Pearsoll Smith) who thought well of the project, promised to supply the funds for a building if the workingmen would organize a club. This was soon effected and, as a preliminary experiment, a hall was hired and furnished, a debating society and a course of familiar lectures opened. A membership of 400 and a crowded hall every evening during the three months the experiment was tried were ample evidence that the movement met with popular favor. During the year a more suitable edifice has been built and is now ready for occupancy. The total cost of the building, including the land and furniture, has been $18,000, of which sum, $5,000 was a donation from Mr. Smith, who also made an advance of $12,000 at five percent interest to be paid at the convenience of the club."

The book carried pictures of the front and rear of the building and a description of the interior which contained, gym, bathrooms, kitchen; clubroom, 40 X 50 feet, and library and reading room containing 2000 volumes. On the third floor, an audience room (theatre) with stage and gallery, dressing rooms, fire-proof scenery and automatic fire-sprinklers; also class and committee rooms.

The dues were $1 per year to any male person 18 years old. There were 1,000 members.

The object of the institute as stated in the constitution was "to offer a counter attraction to drinking saloons and immoral places of resort by establishing an institution which shall offer means of social entertainment, of physical culture and of mental training under right direction and on the general principle of right training leading up to instruction."

CHAPTER 9

# Paperweights and People

JUST three years before Millville changed from township to city status, the Whitall-Tatum Company established a wooden mold department in their Millville plant and it was shortly after this that the more expert glassblowers began experimenting with molten glass. Some of the blowers made paperweights. It has been said that paperweights were made by blowers in the Whitall-Tatum Company's factories from the time Millville became a city until 1912. Some may have been made earlier.

Fancy pitchers, bottles, footed inkwells, and fancy containers were made at the Millville plant.

It wasn't long after the wooden mold department was established that paperweights began to appear. The first ones were crude and solid and couldn't be compared with the later, much more beautiful products. Included among the first ones were the *fountain, devil's fire, swirl* and other plain flat types. It was just about this time that the *Millville Lilly* appeared. The original effort was greatly improved upon because those first ones were poorly made with few colors, which were quite often, poorly selected. It was in the next decade that the finest *Lillies* were made.

The progress and improvement, however, seemed to be rapid and by 1880, many beautiful and unusual paperweights were made at the Whitall-Tatum Company factories. A few years later, John Rhulander, Emil Stanger, Marcus Kuntz, John Fath, Michael Kane and Ralph Barber appeared as glassblowers in the Whitall-Tatum Company's plant and these men were rated as the best in the art of glass blowing. Kane was credited with having created some very unusual paperweights. One of them shows a hunter in a field shooting at a quail with his dog on point and there are two quail perched on a log in the foreground. There is a rail fence in the background. The other weight shows a white dog with orange-brown spots pointing quail and in the background are three trees in pale green.

These works of Kane are very rare and as far as is known, there are none in Millville today.

It is stated by experts, however, that the outstanding paperweight made at the Millville factory was the *Millville Rose* by Ralph Barber. However, some writers have credited Stanger, Kuntz and Rhulander with having made them also.

Barber was several years perfecting his rose paper-

Early Millville Paperweights. *Photos courtesy Earl DuBois.*

weight. They were made in deep rose, pink, white and yellow, with and without stems. The rose is usually upright but some show it tilted and there are a few showing it suspended on its side instead of standing upright. Most of the Barber paperweights rest on a heavy circular foot. Some have a standard with a plain cylindrical stem and a few are known with baluster stem.

According to best available information, Barber's production of paperweights was from 1905 to sometime prior to 1912 when he left the Whitall-Tatum Company employment to become the plant superintendent of the Vineland Flint Glass Works, operated by Victor Durand, of Vineland.

Mr. Barber was well known in Millville where he lived for many years in one half of the former brick residence, now demolished, on the west side of High St., next to City Hall. He helped organize the Millville Social and Athletic Association and was its vice-president for most of its existence. The association occupied the building now the Millville Police Dept.

Emil Stanger and John Rhulander are described as two of the best of Millville's expert glass blowers of all time.

A number of persons who worked during the time these men were employed here were in agreement that there were no better glassworkers ever employed in Millville than those two men.

"They could make most anything," said William McCarthy, who was employed in the same factory when they worked at South Millville.

Stanger was one of the shop that made the 108 gallon bottle, largest ever made.

Paperweights and other fancy mugs, etc. were not manufactured by *Whitall-Tatum Company*. They were produced in the spare time by the workers who were obliged to pay the company for the glass they used. But they sold their product.

Craftsmen in the wooden mould department were obliged to remain on the company premises until 11 a.m. even though they had made their quota of ware by 9 a.m. In other words, the more talented the worker, the sooner he made his quota of ware for the day's work. So the men began dabbling around with the glass, creating all kinds of formations. Some were far better than others.

So, quite a number of the workers earned considerable extra money by making paperweights, etc. in their spare time and then selling the finished product.

Paperweights were very common about Millville around the turn of the century and even the Barber weights could be purchased for $4 to $6.

Besides those skilled craftsmen who have been mentioned these men have been described as being very good: Charles Stratton, Charles Pepper, Benjamin Cox, Wilson Homan, Moses Bailey, John Fath, Horace Rhubart, Tony Stanger, Jacob Hoffner; Clarence Hughes and Charles Reeves must be included as real artists, as cutters and engravers. Clarence S. Reeves, a son, possesses the fine work executed by his father.

The reason that fancy glass products were not made to any extent by these men after 1912 was because of a change in the working conditions and hours and the men had no more spare time in the factory.

Victor Durand, who came to the United States from France, is said to have been employed for a short time at the South Millville plant of Whitall-Tatum Company, then went to Vineland and built a factory where he employed many of Millville's finest glassworkers.

Today Durand Ware is a collector's item.

A former Millville man, perhaps two Millvillians, can be said to be principally responsible for spreading information about the Barber Rose paperweight around. They are William Breeden, now 78, son of the late Mr. and Mrs. Walter Breeden and Edward Griner. Both have been delving into paperweights, antique glass ware and historic merchandise, pictures, etc. for years.

Fact is George and Helen Kearin, who compiled a large book on fancy glass, glass plants and glass manufacturing over the country, obtained information from these two men and Edward W. Minns, another writer who told about paperweights, secured the information he used about Millville weights, including the Barber Rose, from Breeden and Griner.

Breeden's statement is that when the three sat down and talked, it was the general understanding that the article would be published under this by-line, "As told to Edward Minns by William Breeden and Edward Griner." It came out under the name of Minns.

Breeden first met Minns in Newburg, N. Y. and it was there that they had their first talk on paperweights. At the time, the Barber paperweight was not so well known. The price had not reached $500.

Mr. Breeden recalls that, during a trip into North Jersey, he had occasion to visit Newton, a community in which he had once lived. It was near the home of the Dorlinger Glass Company, which created very fine cut glass, at White Mills.

Breeden came upon some paperweights that attracted his curiosity and was told by the lady who owned them they were made by a man by the name of Barber. Puzzled, because he was not aware that Ralph Barber ever lived in that section of North Jersey, he questioned her further and eventually it came out that it was really the father of Ralph Barber who made the weights, who later moved to Millville. And while he never learned for a certainty whether father taught son, to make the kind of weights he did, it is possible.

Emil Larson was another expert maker of paperweights and he made them after Barber was dead. Larson lived in Vineland for years but today he is quite old and has resided in Florida for a number of years. Breeden and William McCarthy described Larson as one of the glass artists of the century.

William Breeden has been delving into paperweights and their history for so many years that he decided he would learn to make them and while he has not yet become a perfectionist, he has made some weights that he says "aren't so bad". At least he knows how to make them.

Arthur Gorham of West Millville, wasn't even a resident of Millville when Barber was alive, but nevertheless, he has made a study of paperweights and is remarkably well informed on the skill and technique required

Early Millville glassblowers at work outside factory building. Approximately 1877. *Virgil Johnson photo.*

to make one. He sells them and gives talks about them. He is without experience in a glass factory but he can give an informative talk on old and modern weights.

Paperweights are still being made and who knows, perhaps, there will be a Breeden weight on sale before long.

According to the Millville Business Directory of 1862, and that was when Millville was still a township and Columbia Avenue was Factory Street and Broad Street was Cinder Road, there were 12 blowers in Millville.

Five lived in Shetterville, four on Mulberry Street, two on High Street and one on Cinder Street.

The railroad was in operation and G. W. Thomas was superintendent of the Millville-Glassboro R.R. That is as far as the rails had been laid. A stage coach, as it was then called, transferred the passengers to Woodbury where they boarded the Woodbury to Camden R.R. Train.

There was a stage coach line running from Millville to Cape Island and Samuel Bishop was manager. George Edward Leeds was postmaster and jeweler on Main Street and Lewis Mulford was president of the Millville Bank as it was then called at Second and Main Streets.

Names carried in the Millville Business Directory of 1862, revealed important employees of the glass factories in that year. They were:

Mr. Lawrence, Mgr., Phoenix Glass Works
Isaac Sharpless, Mgr., Shetterville Glass Works
Thomas Campbell, Mouldmaker, resident Buck Street
Luke West, patternmaker, resident Vine Street
John Bethel, resident Second Street
Joseph H. Headley, packer, Buck Street
John Johnson, resident Factory Street
John Baumgarten, resident Dock Street
Ben J. Van Hook, resident Buck Street
Thomas Corson, carpenter, Shetterville
William Cossaboon, flattenter, Shetterville
Jeremiah Marts, pot maker, Second Street
William Van Hook, bottle maker, Second Street
Joseph Green, carpenter, Whitall Avenue

## *Glassblowers*

A. S. Messer, Second & Mulberry St.
James C. Hassett, Mulberry St.
Henry Miller, Mulberry St.
Sam Miskelly, High St.
Joseph Leight, Cinder Street
Edwin Conover, High Street
Dan Bethel, Mulberry Street
Charles Payne, Shetterville
Anthony Getsinger, Shetterville
Albert Hankins, Shetterville
Isaac Van Hook, Shetterville
Jerimiah Stratton, Shetterville

Other blowers of the era included:

Amos Pennington
"Yeller" Vanaman
"Stiffy" McDonald
"Spikey" Bill Adams
Joe Shields
Chas. Mullen
"Renzo" Suthard
Harry (Trapper) Charlesworth
Billie Surran
"Tackhammer" Kears
"Drum" Pennington
Walter (Reddy) Barber

CHAPTER 10

# Wheaton Industries and Wheaton Village

THE T. C. Wheaton Co. was founded in 1888 by Dr. Theodore C. Wheaton. It is now known as Wheaton Glass Company with Frank H. Wheaton, Sr. Chairman of the Board of Directors and Frank H. Wheaton, Jr. as President of the company. Laur Don Wheaton and other members of the Wheaton family are associated with the company.

The Wheaton Glass Co. is associated with the original firm in the manufacture and production of glass and electronics devices. Both companies are owned by descendants of the founder, Dr. Theodore Wheaton.

From a small beginning with possibly 50 employees in 1888, the Wheaton companies have shown tremendous growth. In fact, the glass firm is probably the fastest growing industry in the community. In the late 30's there were about 350 workers in the plants and today the total number of employees is approximately 3,000.

One of the earliest known views of the original Wheaton glass factory. *Photo courtesy Wheaton Industries.*

The Wheaton plant has out-grown the community and, from an humble beginning, it has reached a point where it is Millville's largest industry. It has an international reputation, sending its products to the far corners of the earth.

Frank Wheaton, Sr. entered the business with his father in 1901, soon after it was established.

The real expansion program of the firm began in 1935 and was instituted by Frank Wheaton, Jr. The new buildings erected by the firm of late are marvels in modern construction and compare favorably with any industrial plant in the nation.

The company owns a plant at Sao Paulo, Brazil, and

provides technical assistance to glass companies of several European countries.

The company's expansive research and business offices on Wheaton Avenue, erected in an attractive setting, also house a stage and assembly rooms, banquet hall, conference rooms and bedrooms. There is another good sized office building on G street.

The Wheaton enterprises are numerous and they are housed in attractive factories in several municipalities.

A large electrically operated glass making plant, one of the few in the nation, has been in operation less than two years. Also included in the later erected structures is a tube shop.

The Wheaton Glass Company has a world-wide trade and Earl DuBois heads that service. He is vice-president of the company's Over-Seas Sales.

The new Wheaton factories have the appearance of college structures. They bear no resemblance to the glass factories in Millville at the turn of the century.

Frank Wheaton, Jr., has proved himself a genius. With his father, now 90 years of age, but quite active, Frank, Jr., is now directing the Wheaton business program and although the many Wheaton enterprises require much of his attention, he finds time to act as director of a bank, an electric company and a hospital.

## *Wheaton Village*

Wheaton Village, a re-creation of a mid-Victorian glass making community, opened to the public on Friday, Oc-

Wheaton Village, a re-creation of a mid-Victorian glass making community, under construction. *Photo courtesy Wheaton Historical Association.*

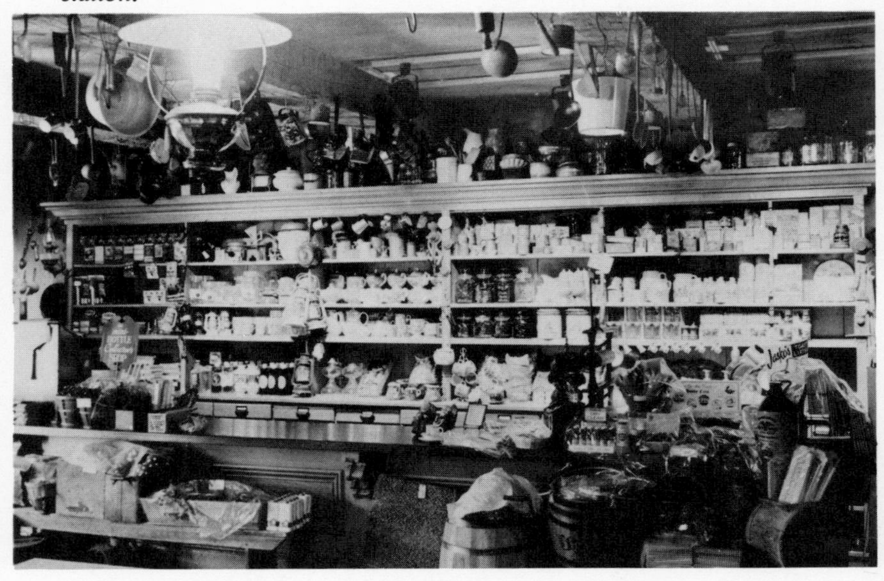

Interior view of the Wheaton Village store. *Photo by Ralph. Courtesy Wheaton Historical Assoc.*

tober 23, 1970. The Village is open daily, except Sundays, from noon until 8 p.m. and Saturdays from 10 a.m. It is located on more than 80 acres of land, paralleled on the east by the new Route 55 and on the west by Glasstown Road.

Wheaton Village boasts of some fifty buildings, authentically reproducing the architecture of the era. The most prominent building in the Village is the Wheaton Museum of Glass. Reminiscent of Victorian Cape May, columned porches and Jersey cedar siding are dominent. The Museum will display an extensive collection of South Jersey crafted glass and will house the finest glass research library related to South Jersey Glass.

The grand opening of Wheaton Village is planned for early Spring, 1971, upon the completion of the glass factory. The glass factory will be patterned after the original 1888 factory started by Dr. T. C. Wheaton. Within the factory itself, spectators will be able to watch the blowing of glass and the hand-crafting of glass products.

Wheaton Village is owned by the Wheaton Historical Association, a non-profit corporation founded for the purpose of preserving the history of, and examples of, South Jersey glass making and other crafts.

CHAPTER 11

# Glass Plants of Yester-Year

IT is possible for a big industry to close down and just fade away, and, with the passing of the years, be forgotten. It seems to have been that way with the Caloris plant, the International Glass Company, and the International Development Company. There may be only a few people in town today who are familiar with the history of these operations, and yet, in the 1914–15 era all were important to the City of Millville.

Most important, perhaps, in the negotiations that brought these concerns into Millville were the late Mayor Walter Felmey who was President of the Millville Municipal League, now the Chamber of Commerce, and City Commissioner W. Fred Ware, a real estate broker, and Paul O. E. Frederick, then managing the Caloris Company's plant in New York City, who was induced to move the works to Millville. The original Caloris Building was eventually acquired by Herbert Sanders, and when it burned, Sanders replaced it with his present structure.

Mr. Frederick was credited with interesting outside capital in the construction of the International Glass Company's plant along the railroad tracks east of the Caloris plant.

H. K. Mulford Company, R. C. Ware Publishing Company, Kolb's Baking Company, Charles E. Hires Company, J. Franklin Meeham of Philadelphia, and Joseph Shoemaker, Bridgeton, were among the men who invested money in the erection of the glass plant which, at one time, employed as many as 1,000 workers.

At the outset, the company acquired a six-pot furnace that had been built some years before by several Millville glassworkers including Charlie and Joe Hankins, Richard, Robert, and Frank Ramsey. It had been idle some time when the International rebuilt and enlarged it. For a time, Paul Frederick was general manager. Bert Beckett, who was at one time plant manager at the upper works of Whitall-Tatum Company also managed the plant for a time but left to direct operation of a plant in Virginia. William McCarthy was, for a time, factory manager.

There was considerable enthusiasm in Millville when the plant started. It was located upon a site bought by contributions from citizens. The project was sponsored by the Municipal League and City Commission.

As a result of the location of the International Glass Company's plant here, there came into being the Inter-

View of International Glass Co., ca. 1915.

COMPANY

national Development Company which built Millville Manor and the homes along Caloris Avenue and that section of Wheaton Avenue. The company gave Millville its first and only golf course with an attractive clubhouse which was later offered to the citizens of the community.

World War I slowed building operations, and after the war came a depression, and thermos bottles were being made in Germany and Japan and sold in the United States cheaper than they could be made here, and International lost some big orders. Conditions failed to improve, and the Philadelphians with the money pulled out and the plant closed down.

George Bornerman got the lamproom business; Nelson Creamer the lubricating oil cup trade, and Paul Frederick took over the tube making business, and later took on a partner, and it became known as the Friedrich-Dimmock Company, and the plant still operates on Main Road at Lincoln Avenue.

Millville Manor is all that is visible of the investment by H. K. Mulford-Ware group, but the remains are still attractive, and the homes are well kept by the owners.

The Millville Bottle Works was founded in 1904 by W. Scott Wheaton, James E. Mitchell and Clarence Corson.

This plant was a conventional pot furnace. There were eight large pots and two small pots. Each shop was made up of eight men and boys.

Glass containers were produced in flint, amber and light green. This factory specialized in private mold ware

—largely pharmaceutical with lettered inserts, perfumes, patent medicine and household needs. A press shop produced stoppers which were individually ground into the necks of perfume bottles.

The Millville Bottle Works operated until 1922. After standing idle for several years the factory and auxiliary buildings were completely destroyed by a series of fires.

CHAPTER 12

# The Making of Glass Containers The Old and the New

AUTOMATION in the glass container industry has left few old-time blow shops in existence. The glass bottle blower, who, at the turn of the century, may have been the highest paid mechanic in the country, is, for the most part, just a memory. There are a few shops in some factories still being operated. The automatic machine has eliminated the jobs of thousands of skilled bottle blowers.

Glassworkers today, live in attractive and comfortable homes, own automobiles, and enjoy the better things in life, a far cry from the whitewashed company houses which most glassworkers occupied a century and more ago.

As has been said, organization came before automation. Today there are two powerful glassworkers' unions,

Current offices of the Kerr Glass Manufacturing Corporation with plant stacks and batch-house showing in the background. This is the oldest continually operating glass manufacturing plant in the United States. Initial operations began in 1806 when James Lee founded the first plant on this site.

one known as the Glass Bottle Blowers' Association of the United States and Canada, and the other, the American Flint Glass Workers' Union, both affiliated with the AFL-CIO.

Today there are two great glass container manufacturing plants in Millville, Wheaton Industries and the Kerr Glass Mfg. Co., employing upwards of 4100 workers.

In nearby Bridgeton and Vineland are more large glass plants, Owens-Illinois in Bridgeton, and the Kimble Division of Owens-Illinois in Vineland.

The basic processes used in forming articles of glass are blowing, pressing, drawing and casting. Each of these may be accomplished either by hand methods or by automatic means. Originally, of course, the hand operations were developed and machinery has evolved only in this century to lighten man's burden or to speed production or both.

Blown glass, which necessarily depended upon the blowpipe, is said to date from the third century B.C., when that useful implement was invented, probably in Egypt or Syria. When glass is blown by hand it may be free-blown or moulded. Free-blown glass may best be understood by visualizing a child blowing a soap bubble. The bubble is not confined and, pursued to its limit, will burst. Glass on the end of a blowpipe behaves in the same manner. If, however, the expansion of the bubble is stopped short of the bursting point it can, depending on the skill of the handler, be fashioned into any of a myriad of sizes and shapes. These can be further adorned with applied decorations of the same or different colored glass.

Moulded glass is made by introducing molten glass on the end of a blowpipe into a cast iron mould, consisting of hinged halves and a bottom plate, then "blowing the bubble" until the glass is forced out tight against the cavity, or inside, of the mould. The inside of the mould may be lettered or otherwise decorated so that the imprint will appear on the finished container. In the case of hand blown bottles, when the above-described operation is completed the bottle is broken away from the blow-

A hand shop at the Shetterville, or South Millville plant of the Whitall-Tatum Co. William H. Cox, finisher; Albert Felmey, blower; William Peek, blower; Ralph Compton, Johnny Baker, Kenneth Hickman, tending boys. *Virgil Johnson photo.*

pipe and, held by a pontil rod affixed to the bottom or a snap-case, the neck of the bottle is reheated and given its final shape, or "finish".

An old time iron-mould shop at either of Whitall-Tatum Company's factories here in Millville was made up of two blowers, a gaffer and three boys: called a "mould boy", a "snapping-up boy" and a "carrying-in boy". The first of these boys sat at or below the level of the factory floor and his function was to close the mould after the blower had got his molten gathering of glass in place and to open the mould so that the blower could remove the bottle when he had finished his operation. The "snapping-up boy" picked up the bottle, using tongs, and placed it in the snap-case which served both as a holder for the reheating operation and an improvised lathe during the turning of the bottle while the gaffer applied the finish. When all these motions had been completed, the "carrying-in boy", using a paddle, carried the bottle to the oven or lehr for annealing.

The whole shop could produce an average of about 220 dozen pint beer bottles for a full day's work. For this the skilled men might even have earned about $8.50. The boys—and these could have been little boys or unskilled men—received $2.76 per week plus tips of about a quarter from each of the skilled workers.

At the beginning of the twentieth century the skills required and the techniques employed in the manufacture of blown glass bottles had changed hardly at all in more than a thousand years.

Pressing glass by mechanical means was an American invention of the late 1820's. This technique evolved from a much older process of forming small items in simple hand presses. Drawing and casting glass had been done at least as far back as the fourteenth century B.C. thus in 1900 glassware made in Millville, as elsewhere, was hand blown, drawn or pressed by the same time-honored methods.

Then in 1903, Michael Owens invented the first fully automatic bottle-blowing machine. Since that date automation in the glass container industry has become almost complete. The glass bottle blower who, at the turn of the century, may have been the highest paid mechanic in the country, is found today in only a few shops in scattered factories.

The modern automatic machine can produce more bottles in half an hour than a hand shop made in a whole day. From a tank capable of melting one hundred tons or more of glass in 24 hours, gobs of molten glass are delivered by gravity to the forming machine at the rate of from 60 to 120 per minute. The moulding of a bottle is accomplished in two steps, of which the first results in the formation of a shape called a "parison" or blank. This is transferred, while still plastic, to the "final" mould of the finished size and shape. These operations are performed by compressed air.

The bottles are placed in the annealing lehr by automatic "hands" and tempered at high speeds and carefully controlled temperatures. About the only thing that hasn't changed is the sorting and packing at the "cold-end" of the lehr.

Although the glass container industry bears little resemblance to that of three quarters of a century ago it is still the economic backbone of Cumberland County. The hundreds of jobs that were lost to automation have developed into thousands of opportunities for today's glassworkers. At the same time life for everyone is a lot easier because of today's higher earnings and the more extensive use of glass as one of the many conveniences of modern living.

CHAPTER 13

# The Largest Glass Bottle in the World

THE largest glass bottle in the world was made at the South Millville plant of Whitall-Tatum Company, now the Kerr Glass Company and was exhibited in the 1903 St. Louis World Fair. It was shattered into a thousand pieces in shipment to another exhibit. There is only a picture left to remind the people of Millville that the world's largest bottle was made in a Millville glass factory by expert glassworkers who are all dead.

Capacity of the bottle was 108 gallons. It was one of 20 or 25 made or partly made. Many of them collapsed in the making. But the determination of Emil Stanger, Marcus Kuntz, John Fath, Sr., and "Tony" Stanger, four artists of the trade, produced a bottle that would hold together.

Fath and Stanger were born in the Alscace-Lorraine area between France and Germany and they, with an-

The largest glass bottle ever blown by hand is shown above with the men who made it. L. to r.: Marcus Kuntz, Tony Stanger, John Fath and Emil Stanger.

other expert glassblower from that country by the name of Ney, were enticed to come to the United States by higher pay.

There were others who worked on that shop and assisted, one way or another, in the manufacture of the world's largest bottle. John Mulford, S. Second Street, had a part and William McCarthy, worked on the next shop and was an observer.

The world's smallest bottle was also made at the same plant by Alex Querns. A single drop of liquid would fill it. There are a number of them around town.

Millville has been a glass manufacturing town since 1806, four years after it was incorporated as a township and time was when there were five glass bottle making plants operating at the same time in Millville, the two Whitall-Tatum plants, Millville Bottle Works, T. C. Wheaton Company and the International Glass Company.

So it could be said that Millville is one of the formost glass manufacturing cities in the nation but, although the largest bottle was made here, the largest piece of glass was made at the Corning Glass plant in Corning, N. Y. It is a 200-inch reflector made for the world's largest telescope, the Haletelescope at Mount Palomar Observatory. The second largest was cast for the same telescope but was never finished. Each weighed 20 tons before being ground down.

# EPILOGUE

TODAY there are two great glass manufacturing operations in Millville, Wheaton Industries and the Kerr Glass Manufacturing Corporation. Wheaton Industries had its beginning in a humble way when several ambitious glassblowers and Dr. Theodore C. Wheaton combined their resources and the local Kerr plant is the outgrowth of a dream of James Lee who first built a small building along the Maurice River at the foot of what is now Mulberry Street and Frank Shetter who built a factory a third of a mile south of what was then the center of the city.

When the first glass factory was established in Millville at the turn of the nineteenth century, it is hardly likely that use of the terms "Men" and "Management" was common practice. However, regardless of descriptions, these men sowed the seed in fertile soil and nourished them well. For years men's labors were backbreaking. The lack of education, property, little community life plus insufficient knowledge of the glass industry took their toll. But these men were of sturdy stock as is evidenced by generations then unborn who fought bravely against hardships of depressions and loss of family ties, in whole or in part during the Civil War, the Spanish American War, World Wars I and II, Korea and Viet Nam. The success of today's glass industry is the crowning glory of those pioneers who envisioned an industry serving all mankind.

"The Big Shop," made the big bottle, 108 gallons, for display at the St. Louis World Fair in 1903.

Seated, left to right: Smith Meyers, tended ovens and carried in; Tony Stanger, born in Millville, a gatherer; John Fath, from Alsace-Lorraine, foot blower; Emil Stanger, gaffner, head of shop, from Alscace-Lorraine; Marcus Kuntz, from Alscace-Lorraine, servitor; Harvey Harris, Millville, sticking-up boy.

Standing, left to right: John Cossaboon, MHS football player, a knocking off boy; Dave Crowley, Millville, mould boy; Charles Stanger, Millville, warming end boy. *Virgil Johnson photo.*

# APPENDIX A

## OLD GLASS WORKS OF NEW JERSEY

1. WISTAR. 1739–circa 1780. Near Alloway, Salem Co.
2. STANGER BROTHERS. Circa 1781. Merged with No. 10 in 1824. Glassboro, Gloucester Co.
3. EAGLE. 1799–circa 1885. Port Elizabeth, Cumberland Co.
4. BROWNS PINES. 1800–? Location unknown. Probably No. 5
5. JONATHAN HAINES. Circa 1800–1825. Clementon, Camden Co.
6. BLOWERS' CO-OPERATIVE. Before 1806. In or near Millville, Cumberland Co.
7. JAMES LEE. 1806– still operated by Armstrong Cork Co., Millville, Cumberland Co.
8. UNION. Prior to 1811–circa 1815. Port Elizabeth, Cumberland Co.
9. COLUMBIA. 1812–prior to 1844. Columbia, Warren Co.
10. HARMONY. 1813–Bought by Owens Bottle in 1918. Still operating. Glassboro, Gloucester County
11. FRANKLIN. 1813–1881. Malaga, Gloucester Co.
12. MARSHALLVILLE. 1814–after 1863. Marshallville, Cape May County.
13. GIDEON SCULL. ?–1820. Millville, Cumberland Co.
14. COFFIN & HAY. 1817–1857. Hammonton, Atlantic Co.
15. WATERFORD WORKS. Circa 1824–1880. Waterford, Camden Co.

16. JERSEY GLASS CO. 1824–early 1860's. Jersey City, Hudson Co.
17. JOHN SCOTT. 1825–1877. Estellville, Atlantic Co.
18. JACKSON. 1827–prior to 1877. Jackson, Camden Co.
19. GREEN BANK-GLOUCESTER LAKE AREA. Dates unknown. Operating in 1828. Possibly at Clark's Landing, Atlantic Co.
20. ATLANTIC. 1830–1862. Nesco, Atlantic Co.
21. NEW BROOKLYN "STANGERS". Two works. 1831–1876. New Brooklyn, Gloucester Co.
22. WINSLOW. 1831–after 1884. Winslow, Camden Co.
23. MULFORD & HAY. 1832–taken over by Whitall-Tatum (No. 7) in 1852. Millville, Cumberland County.
24. NEW FREEDOM. Camden Co. Gordon's History (1834) mentions a works here.
25. STANGERS' "LEWISVILLE". 1834–1893. Glassboro, Gloucester Co.
26. FREE WILL GLASS COMPANY. 1835–combined with "Washington" (No. 31) in 1857. Williamstown, Gloucester Co.
27. DENNISVILLE GLASS MFG. 1836–?. Dennisville, Cape May Co.
28. BRIDGETON. (later Cohansey) 1836–? moved to Penna. in 1900. Bridgeton, Cumberland Co.
29. SOOY & THOMPSON. Prior to 1837. Probably taken over by William Coffin 1840. Green Bank, Burlington County.
30. PENDLETON. 1837–circa 1860. KRESSON, Burlington Co.
31. WASHINGTON. 1839–after 1918. Williamstown, Gloucester Co.
32. WM. COFFIN. 1840–1858. Green Bank, Burlington Co.
33. EXCELSIOR. 1841–1866. Camden, Camden Co.
34. LUMBERTON. Mentioned by Barber & Howe's Recollections (1844). Lumberton, Burlington County

35. WILLIAM PORTER. Early 1840's–circa 1923. Medford, Burlington Co.
36. JESSE RICHARDS. 1848–1868. Batsto, Burlington Co.
37. BODINE & ADAMS. 1848–1885. Tansboro, Burlington, Co.
38. HAY, COFFIN & CO. dates unknown, circa 1850. Camden, Camden Co.
39. FISHER & BECKETT. 1850–1911. Clayton, Gloucester Co.
40. CROWLEY. 1851–circa 1866. Crowleytown (Wharton Tract), Burlington Co.
41. LEBANON. Circa 1851–1866. Lebanon State Forest, Burlington Co.
42. REED & MOULDS. 1854–? Jersey City, Hudson Co.
43. PROGRESS. Circa 1852–1855. Riverside, Burlington Co.
44. ELIZABETH PORT. Circa 1858–? Short-lived. Elizabeth, Union Co.
45. ROWLEY'S BULLTOWN WORK. 1858–1870 (Waldo), Wharton Tract, Burlington Co.
46. ALVA. Dates of founding unknown. Property sold in 1896. Salem, Salem Co.
47. HOLTZ, CLARK & TAYLOR. Early 1860's–circa 1872. Salem, Salem Co.
48. BERGEN POINT. Operated in the 1860's. Bayonne, Hudson Co.
49. JERSEY CITY FLINT GLASS CO. 1862–circa 1880. Jersey City, Hudson Co.
50. JERSEY CITY WINDOW GLASS CO. 1862–circa 1872. Jersey City, Hudson Co.
51. HALL, CRAVEN & PANCOAST. 1862–still operating as Anchor-Hocking, Salem, Salem County.
52. HIRES & COMPANY. 1963–1912 or later. Quinton, Salem Co.
53. R. D. WOOD COMPANY. Circa 1863–1881. Millville, Cumberland Co.

54. WESTVILLE FLINT GLASS. 1865–? Westville, Gloucester Co.
55. RIVERSIDE. 1869–1889. Riverside, Burlington Co.
56. SCOTT & RAPP. 1869–? Short-lived. Herman City (Wharton Tract), Burlington County
57. REFFERS & CO. 1868–before 1872. Camden, Camden Co.
58. W. O. TALCOTT. Circa 1870–before 1872. Atco, Camden Co.
59. R. A. STEELMAN. Circa 1870–before 1872. Estellville, Atlantic Co.
60. FRANK CHEW. Circa 1870 before 1872. Malaga, Gloucester Co.
61. JOSEPH WHARTON. 1870's or 1880's. Camden, Camden Co.
62. MILLVILLE GLASS & MFG. CO. 1871–1901. Millville, Cumberland County
63. F. F. SHARP & CO. ?–1872. Millville, Cumberland Co.
64. F. F. SHARP & CO. ?-1872. Malaga, Cumberland Co.
65. MULFORD & HAY. before 1872–before 1874. Millville, Cumberland Co.
66. BODINE & ADAMS. Probably took over Talcott in 1872 (No. 58)–? Atco, Camden Co.
67. GUESSLING & CO. ?–1872 or later. Jersey City, Hudson Co.
68. GAYNER GLASS CO. Waterford, Camden Co. 1874. Moved to Salem City in 1879. Operating as a Division of Star Glass Co., Salem, Salem County.
69. JOHN FOGER. ?–1877 or later. Glassboro, Gloucester County.
70. GETZINGER AND ALLEN. 1879–after 1889. Bridgeton, Cumberland Co.
71. BENJAMIN LUPTON. 1879–? Bridgeton, Cumberland Co.
72. FISLER & MORGAN. 1880–1915. Clayton, Gloucester Co.

73. JOSEPH A. CLARK & CO. 1880–still operating. Illinois Glass Co. Bridgeton, Cumberland Co.
74. WOODBURY GLASS CO. 1881–1912. Woodbury, Gloucester Co.
75. LIVINGSTON. Probably a revival of the Lebanon Glass Works. Circa 1882–? Lebanon State Forest, Burlington County
76. BERLIN. Camden County.
77. SICKLERVILLE, Camden County
78. SPRING GARDEN, Camden County. 76, 77, 78 are listed in an Industrial Directory of 1883. The compiler doubts their existence.
79. STANDARD. 1882–? Woodbury, Gloucester Co.
80. WOODBURY. Name unknown circa 1883–? Woodbury, Gloucester Co.
81. HAMPTON & BROOKS. 1883–? Bridgeton, Cumberland Co.
82. KIRBY & McBRIDE. 1883–1906. Bridgeton, Cumberland Co.
83. A STOCK COMPANY. 1883–? Bridgeton, Cumberland Co.
84. BUTCHER & WADDINGTON. 1883–1906. Elmer, Salem Co.
85. MORR-JONAS CO. June 1882–absorbed by No. 73 in 1899. Bridgeton, Cumberland Co.
86. JOSEPH PETERS. Probable name of a lantern and chimney glass work operating for some time in the 1880's. Jersey City, Hudson Co.
87. WEST SIDE GLASS CO. Circa 1879–1885 or later. Bridgeton, Cumberland Co.
88. PARKER BROS. 1885–1918. Bridgeton, Cumberland Co.
89. WHITNEY BROS. operating in the 1880's. Camden, Camden Co.
90. ATCO GLASS MFG. CO. 1884–1901. Atco, Camden Co.
91. SWEDESBORO. 1886–1920. Swedesboro, Gloucester Co.

92. PENNS GROVE. A local co-operative venture. 1886–1895 or later. Penns Grove., Salem County.
93. CRYSTAL. 1896–? Camden, Camden Co.
94. T. C. WHEATON CO. 1888–still operating. Millville, Cumberland County
95. VINELAND GLASS CO. 1888–circa 1890. Vineland, Cumberland Co.
96. WALSH BROS. 1890–1905. Janvier, Gloucester Co.
97. BASSETT & CO. 1896–1909. Elmer, Salem Co.
98. WILLIS-MORE. 1890–1909. Fairton, Cumberland Co.
99. AMERICAN BOTTLE CO. 1893–? Vineland, Cumberland Co.
100. GEORGE MESSECK. 1895–? Rosenhayn, Cumberland Co.
101. VICTOR DURAND. 1895–early 1930's. Vineland, Cumberland Co.
102. GEORGE JONAS CO. 1895–about 1920. Minotola, Atlantic Co.
103. TAYLOR & STITES. Circa 1898–1923. Cape May Court House, Cape May County.
104. BEAUMONT GLASS CO. Dates unknown. Bridgeton, Cumberland Co.
105. BRIDGETON GLASS MFG. CO. ?–1906. Bridgeton, Cumberland Co.
106. WACHER. ?–1906. Camden, Camden Co.
107. VINELAND FLINT GLASS CO. ?–1927. Vineland, Cumberland Co.
108. EGG HARBOR CITY. ?–? Probably preceeded the Liberty Cut Glass Co. Egg Harbor City, Atlantic County.
109. GLENCOR GLASS WORKS. Magnolia, Camden Co. Aug. 1900–about 1903.
110. MANUMUSKIN. Cumberland Co. Dates unknown. Mentioned by Dr. Cook, State Geologist in 1868 report as not being in operation.